Cambridge Elements ≡

Elements in the Philosophy of Mathematics
edited by
Penelope Rush
University of Tasmania
Stewart Shapiro
The Ohio State University

INNOVATION AND CERTAINTY

Mark Wilson
University of Pittsburgh

CAMBRIDGE
UNIVERSITY PRESS

CAMBRIDGE
UNIVERSITY PRESS

University Printing House, Cambridge CB2 8BS, United Kingdom

One Liberty Plaza, 20th Floor, New York, NY 10006, USA

477 Williamstown Road, Port Melbourne, VIC 3207, Australia

314–321, 3rd Floor, Plot 3, Splendor Forum, Jasola District Centre,
New Delhi – 110025, India

79 Anson Road, #06–04/06, Singapore 079906

Cambridge University Press is part of the University of Cambridge.

It furthers the University's mission by disseminating knowledge in the pursuit of
education, learning, and research at the highest international levels of excellence.

www.cambridge.org
Information on this title: www.cambridge.org/9781108742290
DOI: 10.1017/9781108592901

First published 2020

A catalogue record for this publication is available from the British Library.

ISBN 978-1-108-74229-0 Paperback
ISSN 2399-2883 (online)
ISSN 2514-3808 (print)

Innovation and Certainty

Elements in the Philosophy of Mathematics

DOI: 10.1017/9781108592901
First published online: December 2020

Mark Wilson
University of Pittsburgh

Author for correspondence: Mark Wilson, mawilson@pitt.edu

Abstract: In *The Vanity of Dogmatizing* (Glanvill 1661, p. 209), the generally skeptical Joseph Glanvill conceded: "And for Mathematical Sciences, he that doubts their certainty, hath need of a dose of Hellebore." But is this familiar opinion truly defensible? Beginning in the nineteenth century, mathematicians began to reposition old problems within revised settings in which formerly accepted truths become overturned. How can mathematics retain its status as a repository of necessary truth in the light of these revisions? This monograph provides a historical sketch of how these challenges have been philosophically addressed.

Keywords: innovation, certainty, Dedekind, De Morgan, setting

ISBNs: 9781108742290 (PB), 9781108592901 (OC)
ISSNs: 2399-2883 (online), 2514-3808 (print)

Contents

Preface

This tract is perhaps written in a more didactic manner than others in the series, but its motivating purpose is largely the same: to acquaint its readers with important aspects of thought within the philosophy of mathematics. I would like to thank Jeremy Avigad, Kathleen Cook, Rachael Driver, Anil Gupta, Julliette Kennedy, Michael Liston, Penelope Maddy, and Stewart Shapiro for helpful comments on these topics.

1 Suggestions from the Symbols

> But now the way seems open to us, still further to generalize the Abstract Geometry, with the help of suggestions arising from the symbols themselves, using the words point, line, etc., in a proper sense consistent therewith. . . . Two questions naturally arise: (1), Is there any geometrical utility in this extension? (2), Is it legitimate to use the postulated properties of the abstract points, lines, etc., in order to prove relations existing among the real points, lines., etc., that is, relations which can be stated without any reference to the abstract elements? . . . [And] it may be said, briefly, that experience has amply shewn that the gain in the generality of the statements of geometrical fact, and the increased power of recognizing the properties of a geometrical figure, enormously outweigh the initial feeling of artificiality and abstractness. . . . [T]he introduction of [extra] elements may well have assisted the constructive faculty [of ingenuity]; that this may happen is, indeed, one of the discoveries of the history of reasoning.
>
> H. F. Baker (1923, pp. 143–44)

Mathematicians and philosophers hope that their proposals will remain unblemished forever: permanent monuments of truth chiseled from the impermeable rock of a priori necessity. They can cheerfully concede that future generations may place their results within wider frameworks that seem more accommodating or practical, for any Gauss or Kant should acknowledge that their specific findings can be fit within conceptual contexts that they had not anticipated. Nevertheless, such authors can remain confident that their original discoveries will remain as unperishable truths even within these enlarged surroundings. "A diamond is forever," runs the old jewelry advertisement, and the same assurance applies to the accumulated gems of mathematical and philosophical discovery. Physicists, biologists, economists, and other human creatures must recognize that everything they propose will be eventually overturned, but mathematicians and philosophers need not harbor comparable fears, as long as they have remained properly diligent in their methodological rigors.

But does this vein of thinking constitute a self-flattering illusion? Jurist Oliver Wendell Holmes Jr. thought so: "[L]ogical method and form flatter that longing for certainty and for repose which is in every human mind. But certainty generally is illusion, and repose is not the destiny of man" (1897, p. 3).

The purpose of this Element is to survey some of the challenges that the natural *enlargements of domain* to which mathematics has been continually subjected pose with respect to this conception of apriorist necessity. (These foundational adjustments are often labeled as "changes in setting.") Operationally, the developmental pressures that prompt these shifts come into play when mathematicians attempt to establish a deductive pathway linking locations A and B and discover that their journey will be greatly facilitated if they are allowed to travel through intermediate locations C, lying outside of the boundaries in which the task was originally posed.

An early example arose as sixteenth-century mathematicians attempted to find real number solutions to cubic equations such as $x^3 + px + q = 0$. (Prizes were awarded to contestants who could produce such answers in the quickest time.) In 1545, through brute symbolic manipulations, G. Cardano arrived at the formula we would write today as:

$$x = \sqrt[3]{-\frac{q}{2} + \sqrt{\frac{q^2}{4} + \frac{p^3}{27}}} + \sqrt[3]{-\frac{q}{2} - \sqrt{\frac{q^2}{4} + \frac{p^3}{27}}}$$

In suitable circumstances, this supplies the desired roots. However, the two cubic roots in Cardano's formula seemingly designate "impossible" (= complex) values in many cases, even if these "impossibilities" eventually cancel out when added together. These "impossible values" comprise examples of the useful, out-of-country locations featured in our geographical analogy. We will survey a range of cases of this sort (a good history is provided by Katz and Parshall 2014.)

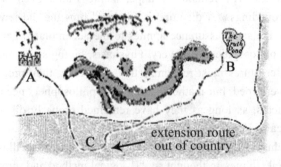

Figure 1. Out-of-country elements

As it happens, most of us have become inured to negative and imaginary numbers from high school algebra, viewing them as unproblematic extracts from "the world of mathematics," without realizing that long into the nineteenth century both items were warily regarded as computational instrumentalities lacking coherent significance. Up to the time of Kant (more or less), the traditional realms of "geometry" and "number" were regarded as the only domains in which the synthetic a priori reasonings of the mathematical sciences can be reliably set. To disarm the present-day complacencies bred by subsequent familiarity, we must first recover the intellectual shock that nineteenth-century observers often expressed when they were first presented with some of these "innovative" changes. For many of us, the strange "extension elements" that mathematicians added to regular Euclidean geometry in the first part of the nineteenth century[1] can still serve this purpose if we have not studied the modern subject of "algebraic geometry," in which these novel ingredients are now central ingredients. Absent such a background, the resulting claims will likely strike many of us as bizarre, viz. the proposition that two apparently non-overlapping circles secretly intersect in four points: two of them regular imaginary and two additional "circular points" on the line at infinity. (We will review the motivating rationale for these weird claims in a moment.) In 1883, one of the prominent developers of these dark arts, the mathematician Arthur Cayley, called for a philosophical examination of their rationale:

> [T]he notion which is really the fundamental one (and I cannot too strongly emphasize the assertion) underlying and pervading the whole of modern analysis and geometry [is] that of imaginary magnitude in analysis and of imaginary . . . points and figures in geometry. This [topic] has not been, so far as I am aware, a subject of philosophical discussion or inquiry . . . [E]ven [if our final] conclusion were that the notion belongs to mere technical mathematics, or has reference to nonentities in regard to which no science is possible, still it seems to me that as a subject of philosophical discussion the notion ought not to be this ignored; it should at least be shown that there is a right to ignore it. (1889, p. 433)

And the answer Cayley himself suggests sounds disconcertedly mystical in its invocation of Plato's cave:

> That we cannot "conceive" [of "purely imaginary objects"] depends on the meaning which we attach to the word conceive. I would myself say that the purely imaginary objects are the only realities, the ὄντως ὄντα ("the realities that really exist"), in regard to which the corresponding physical objects are

---

[1] This period is frequently labeled as the "projective geometry revolution." It should not be confused with the non-Euclidean geometry that became popular later, which actually raises fewer methodological puzzles in its wake.

as the shadows in the cave; and it is only by means of them that we are able to deny the existence of a corresponding physical object; if there is no conception of straightness, then it is meaningless to deny the existence of a perfectly straight line. (1899, p. 433)

Soon thereafter, a range of contemporaneous philosophers (e.g., Ernst Cassirer and Bertrand Russell) actively engaged with Cayley's concerns, often in relatively unsatisfactory ways. But independently of their academic proposals, virtually every working mathematician of the late nineteenth century needed to ponder these methodological issues in some manner or other, if only to realign their own investigative compasses along the axes of fruitful inquiry that were dramatically restructuring the subjects in which they worked. In this Element, we will particularly focus on the intriguing methodological suggestions found in the pithy remarks offered on these topics by the great nineteenth-century algebraist Richard Dedekind, whose methodological shadow has loomed over mathematical practice ever since.[2] Some of his central themes have been largely overlooked by his modern admirers, despite the fact that they paint a portrait of the mathematical enterprise that remains entirely pertinent to our own era – or so this brief tract will argue.

C. F. Gauss designated mathematics as "the queen of the sciences" (displacing theology from its former pride of place), and by equal rights, the *philosophy* of mathematics ought to perch upon a comparable throne within philosophy as well. And that was the prestige with which philosophers of earlier times accorded its methodological concerns. Today, however, the subject has lost much of its former allure, and academic consideration has largely thinned into wan disquisitions on "naturalism" and "ontological commitment." As a result, the puzzles of innovative practices have become relegated to the sidelines of specialized concern, bearing little anticipated relevance to the central concerns of language, metaphysics, or the wider stretches of science. This Element will argue that this demotion is a mistake; an adequate appreciation of the motivational factors that drive mathematics to continually reshape old domains into considerably altered configurations ought to remain a central ingredient within our attempts to gauge the conceptual capabilities of human thought more generally.

Working mathematicians, of course, cannot afford to ignore the reconstitutive adjustments that continually redirect their disciplines in unexpected directions, for their academic standing may depend upon their ability to convince their colleagues that their innovations represent "the right way to proceed." However, a distaste for the disputes about "abstract objects" and so forth that dominate

---

[2] Emmy Noether: "Es steht alles schon bei Dedekind."

current philosophical discussion has induced a profound *horror philosophiae* within mathematical circles, which frequently invoke simple formalist excuses (the "if-thenism" of Section 5) that allow them to beg off "waxing philosophical" in a manner they distrust.

Unfortunately, the considerations that guide research within modern mathematics have become forbiddingly technical, and an adequate mastery of their motivating threads is hard to obtain. To evade these pedagogical obstacles, this Element will largely concentrate on an assortment of easier-to-explain nineteenth-century adjustments in which the winds of innovation altered traditional mathematical landscapes substantially. An excellent starting point lies with those funny points of non-intersecting "intersection" mentioned earlier (which also represents one of the central cases that Cayley worried about).

The main impulse came from algebra. Descartes's innovations within what we now call "Cartesian geometry" forged unexpected pathways between theorems that could not be obtained through traditional Euclidean proof techniques. For example, ellipses, parabolas, and hyperbolas strike us as rather similar in their animating behaviors, but the Euclidean proofs required to establish that these facts differ significantly. In contrast, the same relationships can be established within Cartesian geometry by calculations directed to their common equation $Ax^2 + Bxy + Cy^2 + Dx + Ey + F = 0$, if we are allowed to factor this expression into expressions that lack any obvious significance. But how can we trust a proof if we do not understand what its intermediate steps mean? These considerations prompted synthetic geometers such as J.-V. Poncelet to wonder if similar (yet intelligible) pathways of easy reasoning could not be established as proper to geometry if its internal dominions were extended through defensible policies for extending a preexistent domain. Indeed, supplementary "points at infinity" had already become familiar as the "horizons" and "vanishing points" within a perspective drawing.

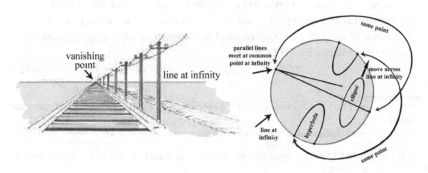

Figure 2. Points at infinity

Poncelet realized that these supplementary objects could be harnessed to significant inferential advantage if we allow ourselves to reason about these "lines and points at infinity" as if they represented regular Euclidean ingredients. From this extended point of view, the mysteriously similar behaviors of ellipses and hyperbolas can be explicated by simply dragging an ellipse across the line at infinity until it appears as if it has become split into two pieces.[3] By allowing parallel lines to intersect at such "points of infinity," we can likewise avoid annoying distinctions in our proofs between lines that cross somewhere and those that do not.

Allied pathways of improved reasoning similarly rationalize the strange "imaginary points" mentioned previously, for reasons that I will sketch shortly. But do we not risk spoiling the a priori certainties of the traditional Euclidean realm by rashly introducing these "extension element" supplements? It would likewise make algebraic calculations simpler if we could assign a factor such as "6/0" a convenient numerical value, but it proves impossible to do so without opening a door to harmful contradictions (i.e., the mere acceptance that "6/0" possesses a value immediately allows one to prove that "0 = 1"). Poncelet plainly requires some more sophisticated form of methodological justification for his innovations than the crudely pragmatic "They allow me to reach nice results quickly." Over the course of this Element, we will examine a succession of proposals to this purpose. We can begin with Poncelet's own justificatory offering, based upon a principle that he dubbed "the permanence of mathematical relations" (other authors call it "persistence of form"):

> Is it not evident that if, keeping the same given things, one can vary the primitive figure by insensible degrees by imposing on certain parts of the figure a continuous but otherwise arbitrary movement, is it not evident that the properties and relations found for the first system, remain applicable to successive states of the system, provided always that one has regard for certain particular modifications that may intervene, as when certain quantities vanish or change their sense or sign, etc., modifications which it will always be easy to recognize a priori and by infallible rules? ... Now this principle, regarded as an axiom by the wisest mathematicians, one can call the principle or law of continuity for mathematical relationships involving abstract and depicted magnitudes. (1822, p. 19)

---

[3] Florian Cajori (1919, p. 62) characterizes the disadvantages of traditional Euclidean geometry as follows:

The principal characteristics of the ancient geometry are:

(1) A wonderful clearness and definiteness of its concepts and an almost perfect logical rigor of its conclusions.

(2) A complete want of general principles and methods.

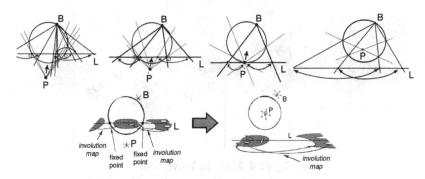

Figure 3. Pole and polar movement

Here is how these notions operate in the context of our "imaginary" geometrical points. Projective geometry asks how images adjust when an originating slide (say, a picture of a cat's head inscribed on the sides of a sphere C) is projected onto varying screens. As this occurs, the ears, mouth, and so forth will distort considerably as the image plane L is moved to different positions relative to a lamp B. However, certain abstract characteristics of the image must remain preserved within all of these placements. (Otherwise, we would not be able to recognize that "it's the same cat" throughout.) The projective geometers discovered that this "invariant" could be explained in terms of a geometrical relationship called a "cross-ratio." In Figure 3, I have tried to illustrate this construct in the upper sequence of diagrams, although the exact details are not important for our purposes.

We carve out a two-dimensional slice of our arrangements and consider how the projected image appears on our "screen" (the line L). By allowing light to travel backward across the line at infinity (!), two projected cat images will always appear on L. When a complete cat head fits inside the circle C connecting to our lamp B, we can locate an important exterior point P called the *pole* of the construction by drawing tangents from the two places where C intersects L (which is then called the *polar* of P). We can then correlate the respective parts of our two cat heads by a suitable mapping **m** called an *involution*. The two positions where L intersects C constitute *fixed points* of **m** in the sense that **m** maps these positions to themselves as self-corresponding. The cross-ratio invariant we seek can then be explicated in terms of the invariant manner in which the correlated cat features cluster together around these two fixed points. (This pattern represents a generalization of a "harmonic division" within traditional geometry.) As a result, the two fixed points represent the central "controlling points" around which the rest of the construction arranges itself.

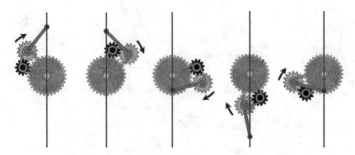

Figure 4. Machine movement

What happens if we now push the pole point P toward the interior of C? In particular, what happens to the polar line L? It will gradually move outward until it passes through a transition stage where L is tangent to C and its two fixed points coincide. Pushing P fully inside C, we reach an altered scenario in which the polar line L now lies outside of C, yet a similar involution pairing **m** between the cat parts remains well-defined (i.e., their cross-ratio remains the same). But **m**'s controlling fixed points have apparently vanished from the scene.

Or have they merely become invisible? Here's where Poncelet's "permanence of relationships" principle enters the story. Examine our successive P/L drawings as if they comprise successive frames within a motion picture film. The resulting montage supplies an evolving picture of integrated movement, much like we would witness in a film of an actual cycling mechanism, such as the Cardan gearing illustrated in Figure 4.

We previously noted that the original fixed points of the mapping **m** gradually move closer together until they fuse and seemingly disappear. Poncelet's principle contends that this "disappearance" is only apparent, because all of the other pole/polar/involution relationships remain intact, albeit altered in appearance. Our two fixed points have merely become "imaginary" by moving out of the plane of our paper, explaining why they can no longer be readily pictured within a conventional visual representation. Nonetheless, these imaginary fixed points continue to "control" the rest of the associated pattern (operating "from an astral plane" as it were). These same supplementary points also supply the mysterious "intersections" between non-overlapping circles to which I earlier alluded. However, a proper justification of their utility in Poncelet's manner requires consideration of the "persisting relationships" evident within a sequence of machine-like movements of the type illustrated.

By successively enlarging the dominions of traditional geometry in this manner, nineteenth-century geometers felt that they had quasi-inductively stumbled onto some hidden Platonic reality that renders the surface relationships of traditional geometry coherent. In 1832, the German geometer Jakob Steiner declared:

> The present work contains the final results of a prolonged search for fundamental spatial properties which contain the germ of all theorems, porisms and problems of geometry so freely bequeathed to us by the past and the present... It must be possible to find for this host of disconnected properties a leading thread and common root that would give us a comprehensive and clear overview of the theorems and better insight into their distinguishing features and mutual relationships. .... By a proper appropriation of a few fundamental relations one becomes master of the whole subject; order takes the place of chaos, one beholds how all parts fit naturally into each other, and arrange themselves serially in the most beautiful order, and how related parts combine into well-defined groups. In this manner one arrives, as it were, at the elements, which nature herself employs in order to endow figures with numberless properties with the utmost economy and simplicity. (1832, p. 315)

In 1857 the president of Harvard, Thomas Hill, rhapsodized similarly:

> The conception of the inconceivable [imaginary], this measurement of what not only does not, but cannot exist, is one of the finest achievements of the human intellect. No one can deny that such imaginings are indeed imaginary. But they lead to results grander than any which flow from the imagination of the poet. The imaginary calculus is one of the master keys to physical science. These realms of the inconceivable afford in many places our only mode of passage to the domains of positive knowledge. Light itself lay in darkness until this imaginary calculus threw light upon light. And in all modern researches into electricity, magnetism, and heat, and other subtle physical inquiries, these are the most powerful instruments. (1857, p. 265)

And these same advantages accrue to the even stranger schemes and divisors characteristic of modern algebraic geometry:

> Our examples show the surprisingly wide range of possible behavior ..., and the apparent jungle of possibilities leads to a basic question: Where are the nice theorems?
>
> A fundamental truth [then] emerges: to get nice theorems, algebraic curves must be given enough living space. For example, important things can happen at infinity, and points at infinity are beyond the reach of the real plane. We use a squeezing formula to shrink the entire plane down to a disk, allowing us to view everything in it. This picture leads to adjoining points at infinity, and in one stroke all sorts of exceptions then melt away. We [will] enhance the reader's intuition through pictures showing what some everyday curves look like after squeezing them into a disk.

Continuing [our] quest for nice theorems, [we] once again [find that] the answer lies in giving algebraic curves additional living space – in this case we expand from the real numbers to the complex. Working over them, together with points added at infinity, we arrive at one of the major highlights of the book, Bézout's theorem. This is one of the most underappreciated theorems in mathematics, and it represents an outstandingly beautiful generalization of the Fundamental Theorem of Algebra. (Kendig 2010, p. ix)

On the other hand, a large number of later mathematicians were troubled by the mystical forms of Platonic appeal that such assertions invoke:

[Otto Hesse regarded Steiner's later period] as marked by his struggle with the imaginary, or as Steiner liked to say, his quest to seek out those "ghosts" that hide their truths in a strange geometrical netherworld.[4] (Rowe 2018, p. 64)

In later sections, we will review how subsequent methodologists attempted to convert Poncelet's worthy (yet dodgy) appeals to the "persistence of relationships" into more rigorous forms of methodological justification.

But we should also observe that such "changes in setting" are not completely irrevocable. Felix Klein comments upon the worthy topics left behind:

In one respect, of course, the Plücker formulas, in spite of their great generality, do leave some problems open: they yield nothing about the separation of the real from the imaginary. Even though abstract thought was indifferent to these questions for decades, they are still of the greatest interest to those who seek the true geometric shape of the varieties. It must be regarded as an aberration of modern geometry that the importance of this question is everywhere denied. . . . Plücker was no "projectivist" in the true sense. In the style of the old geometers of the 18th century he clung to the concrete, investigating [matters] . . . all of whose significance vanishes from the purely projective viewpoint.[5] (1926, p. 114)

Indeed, those old problems have reemerged within the modern context of robotics, in which we need to compute the locations where an automaton is likely to bump into the tables and chairs around it. Learning about their imaginary intersection points in the projective manner is not very helpful in this context. (More correctly, the projective extension elements may still prove useful in these pursuits, but only as halfway houses to the results we really need.)

As remarked earlier, most of us now regard the employment of complex and negative numbers within algebra as "old hat" and not particularly demanding of

---

[4] Compare (Klein 1908, p. 187): "To Steiner, imaginary quantities were ghosts, which made their effect felt in some way from a higher world without our being able to gain a clear notion of their existence."

[5] The formulas cited do not distinguish the real points of intersection between two figures from their "imaginary" crossings.

philosophical attention. But such acceptance is of relatively recent vintage and generally relies upon the Argand diagram and free appeals to equivalence class constructions (whose origins we will review in Section 6). And the historical trepidations associated with these novel ingredients were not ill-founded, for unscrutinized complacency with respect to complex arithmetic can lead to unpleasant surprises. Many readers will perhaps be startled by this simple computational paradox:

$$+2 = \sqrt{4} = \sqrt{(-2)(-2)} = \sqrt{(-1)(2)(-1)(2)} = \sqrt{(-1)^2\, 2^2} = -1 \cdot 2 = -2$$

Something has surely gone wrong, but what is it? A modern response contends that the square root function, properly speaking, "lives" upon a so-called Riemann surface, which (in this case) resembles a two-floored parking lot where we can drive around and around continuously, shifting the local value of $\sqrt{4}$ from +2 to –2, depending upon which floor we reside. If we are unaware of this twisted topology, we remain susceptible to paradoxes of the type illustrated (a professional mathematician is aware of the "branch cuts" tactics required to evade these computational hazards, but many of us never advance so far in our training). Falsely confident, we find ourselves in a situation of a hiker in an enchanted wood who keeps returning to the same spot where everything that formerly seemed right-handed is now reversed mirrorwise. When we clear away the obscuring foliage, we find that we have been walking on a Riemann surface.

In this corrective spirit, Hermann Weyl writes in a famous passage:

> I shared [with Felix Klein] his conviction that Riemann surfaces are not merely a device for visualizing the many-valuedness of analytic functions, but rather an indispensable essential component of the theory; not a supplement, more or less artificially distilled from the functions, but their native land, the only soil in which the functions grow and thrive. (1955, p. vii)

In Weyl's invocation of a "natural soil," we encounter the notion of what mathematicians call "the proper setting" of a mathematical inquiry – that is, the availability of enough distinctions to supply a collection of reasoning with an adequate allotment of what Kendig calls "living space." Elsewhere he recasts

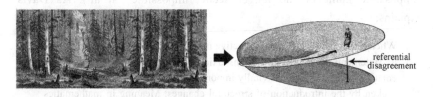

referential disagreement

Figure 5. Living on a Riemann surface

Steiner's conception of a "hidden geometry" as a similar appeal to the "proper setting" for its reasonings: "Hence it turns out that what we see in R^n is just the tip of an iceberg – a rather unrepresentative slice of the variety at that – whose 'true' life, from the algebraic geometer's viewpoint, is lived in $P^n(C)$ [= n-dimensional projective space]" (1983, p. 164).

But we can scarcely discern these "proper settings" through armchair rumination alone; we must quasi-experimentally probe the computational advantages and pitfalls of an unfamiliar landscape in the manner of a cautious recognizance crew. Indeed, the greatest virtuoso of these trial-and-error tactics, Leonhard Euler, describes his methodology for dealing with the hazards of "speculative" exploration as follows:

> If one looks at mathematical speculations from the point of view of utility, they can be divided into two classes: first, those which are of advantage to ordinary life and other sciences, and the value of which is accordingly measured by the amount of that advantage. The other class comprises speculations which, without any direct advantage, are nevertheless valuable because they tend to enlarge the boundaries of analysis and to exercise the powers of the mind. Inasmuch as many researches which promise to be of great use have to be given up owing to the inadequacy of analysis, those speculations are of no little value which promise to extend the province of analysis. Such seems to be the nature of observations which are usually made or found a posteriori, but which have little or no chance of being discovered a priori. Having once been established as correct, methods more easily present themselves which lead up to them, and there is no doubt that through the search for such methods the domain of analysis may be considerably extended. (Euler 1761, p. 657)

Prima facie, these often severe alterations in mathematical settings pose difficulties for the traditional presumptions of mathematical certainty with which this essay begins. Indeed, Philip Davis raises these exact concerns in a survey of how "changes in setting" have overturned presumptive truth-values within many branches of mathematics. He particularly focuses on states of affairs that were once dismissed as impossible or incoherent but have become regarded as cherished landmarks after some innovative "adjustment in setting" has become accepted (to us moderns, Cardano's formerly "impossible numbers" no longer seem "impossible" at all). As Davis explains:

> When placed within abstract deductive mathematical structures, impossibility statements are simply definitions, axioms or theorems. ... [But] in mathematics there is a long and vitally important record of impossibilities being broken by the introduction of structural changes. Meaning in mathematics derives not from naked symbols but from the relationship between these

symbols and the exterior world. . . . Insofar as structures are added to primitive ideas to make them precise, flexibility is lost in the process. In a number of ways, then, the closer one comes to an assertion of an absolute "no," the less is the meaning that can be assigned to this "no." (1987, pp. 176–77)

Although Davis here alludes to "the relationship between these symbols and the exterior world," the judgmental factors that may motivate a significant "change in setting" do not derive from any quasi-empiricism of an Eulerian stripe but stem from a "purely mathematical" concern to isolate the secret structural affinities that link together seemingly disparate areas of human thinking. We will discuss these issues further in Section 7.

In any case, here is the central methodological dilemma that this book will address. To obtain "nicer theorems," mathematicians often alter settings to gain the "extra room" of maneuver that extension element innovations facilitate. But how can such tactics retain mathematics' traditional claims to indubitable certainty if we can permissibly invoke Lands of Cockaigne in which desirable results run around with little messages on their backs, "Take me"? A true believer such as Jakob Steiner would have retorted that undisciplined appeals to novel ingredients would not have been adequately "organically generated" in a "permanence of relations" manner from what came before (you cannot always take what you want). Indeed, we will encounter similar appeals to "organic generation" many times in the sequel. But what on earth can the phrase signify?

Paraphrasing Hermann Weyl once again, *innovation* and *certainty* continually wrestle for the soul of the mathematician in the proverbial manner of devil and angel (1939, p. 500). Our task is to understand what sorts of reconciliation between these extremes are philosophically feasible.

Figure 6. Angel and demon

2 Innovation and Error

Everything that we call Invention or Discovery in the higher sense of the word is the serious exercise and activity of an original feeling for truth, which, after a long course of silent cultivation, suddenly flashes out into fruitful knowledge. It is a revelation working from within on the outer world, and lets a man feel that he is made in the image of God. It is a synthesis of World and Mind, giving the most blessed assurance of the eternal harmony of things.

J. W. Goethe (1906, p. 94)

Frequently the innovative insights that lead to Poncelet-like revisions of prior doctrine emerge in an immediate burst of "Aha! Now I see" insight. Carlo Cellucci comments:

A basic characteristic of Romanticism was the claim of an absolutely free creativity of the mind and its power to solve all problems. The impact of Romanticism on mathematics is apparent from the assertion of several mathematicians in the second half of the nineteenth century, that mathematical concepts and axioms are absolutely free creations of the human mind, subject only to the requirement of consistency. (2017, p. 169)

In an allied vein, nineteenth-century methodologist Franz Reuleaux was impressed by the fact that a "mechanical genius" such as James Watt could repeatedly devise inventions that are optional for the tasks for which they were intended, although Watt could not articulate the unconscious thought processes that led to these discoveries. Reuleaux writes:

A more or less logical process of thought is included in every invention. The less visible this is from outside, the higher stands our admiration of the inventor, – who earns also the more recognition the less the aiding and connecting links of thought have been worked out readily to his hand. (1876, p. 231)

But repeated successes of this character surely indicate that some general form of search procedure resides within these subterranean cogitations, which an able analysist should be able to elevate into explicit articulation. Diagnostic tasks of this nature were characterized as "converting an inventive Art into a proper Science" within the framework of nineteenth century Romanticism. Reuleaux again indicates:

So long as [the science of mechanism] could not reach to the elements and mechanisms of machines without the aid of invention, present or past, it could not pretend to the character of a science, it was strictly speaking mere empiricism – (sometimes of a very primitive kind), – appearing in garments borrowed from other sciences. (p. 20)

In 1844, Hermann Grassmann likewise complains that the production of traditional Euclidean proofs relies excessively upon irregular creative activities that should submit to an improved form of systematization that he likewise dubs as a "true Science":

> Demonstrations are frequently met with, where, unless the theorems were stated above them, one could never originally know what they were going to lead to; here, after one has followed every step, blindly and at haphazard, and ere one is aware of it, be at last suddenly arrives at the truth to be proved. A demonstration of this sort leaves, perhaps, nothing more to be desired in point of rigidity. But scientific it certainly is not. Ubersichtlichkeit, the power of survey, to lacking. A person, therefore, that goes through such a demonstration, does not obtain an untrammeled cognizance of the truth, but he remains – unless he afterwards, himself, acquires that survey – in entire dependence upon the particular method by which the truth was reached. And this feeling of constraint, which is at any rate present during the act of reception, is very oppressive for him who is wont to think independently and unimpededly, and who is accustomed to make his own by active self-effort all that he receives. (1844, p. xxxi)

Generally speaking, nineteenth-century primers devoted to what they characterize as "logic" rarely concentrate upon the first-order inferential relationships central within a modern university course on that subject, but instead concern themselves with the steps required to convert an intuitively established Art into a properly systematized Science. In the next section, we find that when writers like Dedekind identify themselves as following "logic," their intent is often quite different from what a similar invocation of "logic" would mean in the writings of an author from the twentieth century.

Innovative insight is truly admirable, but even geniuses make mistakes! The creative observations that connect topic A to topic B in insightful ways may also skip over transitions which demand greater scrutiny. Let us quickly review an historical crisis of this type due to Bernhard Riemann, whose consequent ramifications play a central motivational role in the pages ahead. (Riemann's lapse in sound reasoning probably comprises the most important of the "crises in rigor" that confronted the nineteenth century mathematicians.[6])

In the previous section, we briefly mentioned Riemann's peculiar "surfaces," which he employed to reorganize our understanding of complex functions in a startlingly efficient manner. But to establish key features of these supportive entities, Riemann relies upon an existence criterion that he dubs as "the

---

[6] Many philosophers presume that this "crisis" centered upon Weierstrass's refurbishing of the calculus in δ/ε terms, an inaccurate gloss that significantly underestimates the difficulties of the actual problems.

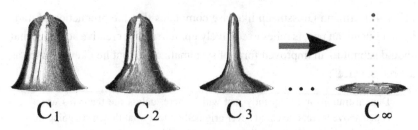

$$C_1 \qquad C_2 \qquad C_3 \qquad \cdots \qquad C_\infty$$

Figure 7. Failure of the Dirichlet principle

Dirichlet principle": if a collection of functions can be graded by positive number assignments, then some minimal function must exist within this set (Monna 1975, Bottazzini and Gray 2013). Figure 7 offers a simple illustration of what can go wrong.

Take a wire rim of arbitrary shape, and apply a soap film to it. Such a membrane stores internal energy according to its degree of bending, so a calculation of the total energy stored within a particular membrane configuration (expressed as an integral over its surface) can grade a proposed membrane as to whether its total stored energy represents the minimum possible for the rim in question. We intuitively anticipate that such a film will eventually subside to an equilibrium configuration of minimal energy (possibly several distinct shapes can accomplish this equally well). These energy measures supply the "positive number" grades required by the Dirichlet principle, which simply converts our physical expectations into a mathematical existence principle. However, Karl Weierstrass and others quickly showed that such a conclusion cannot be universally accepted. Let our "rim" consist of a regular oval *plus* a single point elevated above its center and consider the sequence of bell-like patterns illustrated, where our film attaches to our oval and point in the manner required. As we progressively examine the sequence of shapes C_1, C_2, ..., we find that their total degree of bending continuously decreases yet never reaches a minimum. Their intuitive limit C_∞ contains a discontinuous jump under the elevated point that disqualifies C_∞ from qualifying as a solution to our soap film equation altogether. Accordingly, we are confronted with a descending collection of positive energy films whose lower bound does not represent a mathematical object of the same type as the C_i, contrary to our intuitive presumptions. Without some deep repair, brute appeals to the Dirichlet principle cannot be regarded as reliable and the magnificent edifice that Riemann erected upon its basis stands in jeopardy. Felix Klein characterizes its aftermath as follows:

> The majority of mathematicians turned away from Riemann; they had no confidence in the existence theorems, which Weierstrass's critique had robbed of their mathematical supports. They sought to salvage their

investigations of algebraic functions and their integrals by again proceeding from the given equation. . . . In this respect we can cite a characteristic statement from Brill and Noether's great "The Development of the Theory of Algebraic Functions Past and Present": "In such generality the concept of a function, volatile and ungraspable, no longer admits verifiable deductions." Thus Riemann's central existence theorem for algebraic functions on a given Riemann surface fell from its place, leaving only a vacuum.

Riemann had a quite different opinion. He fully recognized the justice and correctness of Weierstrass's critique; but he said, as Weierstrass once told me: "that he appealed to the Dirichlet Principle only as a convenient tool that was right at hand, and that his existence theorems are still correct." Weierstrass probably agreed with this opinion. For he induced his student H.A. Schwarz to undertake a thorough study of Riemann's existence theorems and to seek other proofs for them, in which Schwarz succeeded. . . .

The physicists took yet another position: they rejected Weierstrass's critique. Helmholtz, whom I once asked about this, told me: "For us physicists the Dirichlet Principle remains a proof." Thus he evidently distinguishes between proofs for mathematicians and for physicists; in any case it is a general fact that physicists are little troubled by the fine points of mathematics – for them the "evidence" is sufficient. Thus, though Weierstrass later proved that there are continuous functions without derivatives, a prominent representative of mathematical physics even today teaches: "It is a law of thought that every function becomes linear in the infinitely small"! I have therefore taken pains to discover the basis for the physicists' attitude toward mathematical rigor. . . .

[T]his basis lies in the fact that the physicists' mathematical thinking, as a consequence of a one-sided habit, is solely in terms of approximations; i.e., it is exact only to within a temporarily given bounded number of decimals. The physical point is a sort of blob, the curve a band. I have been happy to adduce the divergent opinions on the critique of the Dirichlet Principle because it is so apposite to this lecture to show how slow mathematical ideas are to make their way through. (1926, pp. 246–48)

In the sequel, we will return to this paradigm example of a truly "creative departure from mathematical rigor" on several occasions. (Riemann's failure served as a beacon warning for the methodologists like Dedekind who came along afterward.) At present, let me offer a few clarificatory remarks. First of all, physicists often reach deductive conclusions through reliance upon what they regard as "Nature's own solution" to a particular mathematical question:

Enough, however, has been said to show, first, our utter ignorance as to the true and complete solution of any physical question by the only perfect method, that of the consideration of the circumstances which affect the motion of every portion, separately, of each body concerned; and, second, the practically sufficient manner in which practical questions may be attacked by limiting their generality, the limitations introduced being themselves deduced from experience, and being therefore Nature's own solution (to a less or greater degree of accuracy) of the infinite additional number of equations by which we should otherwise have been encumbered. (Thomson and Tait 1912, p. 3)

However, the present situation does not fit this pattern, despite common assertions to the contrary. The equations we (tacitly) utilized for our soap film fail to include any mechanism for frictional damping. A membrane that actually obeys our simplified modeling requirements will wiggle forever and never come to rest. So "Nature's own solution" to the mathematical problem posed will represent a membrane that never comes to rest, not the energetically minimized state augured by the Dirichlet principle.

That being said, certain physicists (e.g., Richard Feynman) commonly dismiss the stern requirements of "mathematical rigor" as unwanted shackles upon their powers of "physical intuition." ("I know basically what's going to happen, so I'm not going to wait until the mathematicians get their i's and t's dotted and crossed to their own satisfaction.") To such contentions, scrupulous mathematicians such as Karl Jacobi have long objected:

> It is true that M. Fourier was of the opinion that the principal aim of mathematics was its public utility and the explanation of natural phenomena; but a philosopher such as himself should have known that the unique aim of science is the honor of the human spirit, and that in this regard, a question concerning numbers is worth as much as a question about the system of the world. (1830, p. 83)

Later in the book, we will examine a striking exemplar of the "unwanted shackles of mathematical rigor" with respect to the so-called Heaviside operational calculus, but, for the time being, let us note that similar finicky details involving Dirichlet-like limiting curves continue to complicate reliable procedure within modern engineering contexts in very practical ways. We will discuss some of these issues later.

However, Klein's remarks raise another important issue with which we should also be concerned. The soap rim counterexample we supplied arises in circumstances somewhat outside Riemann's intended domain of application, and all of the theorems he reached via his faith in the Dirichlet principle have been fully vindicated subsequently. Why should a counterexample arising outside of an intended domain **D** be regarded as a challenge to the rigor of ones reasoning within **D**? This question suggests a softer manner in which we might parse some of Ludwig Wittgenstein's seemingly cavalier queries within his *Remarks on the Foundations of Mathematics*:

> We shall see contradiction in a quite different light if we look at its occurrence and its consequences as it were anthropologically – and when we look at it

with a mathematician's exasperation. That is to say, we shall look at it differently, if we try merely to describe how the contradiction influences language-games, and if we look at it from the point of view of the mathematical law-giver.

But wait – isn't it clear that no one wants to reach a contradiction? And so that if you shew someone the possibility of a contradiction, he will do everything to make such a thing impossible? (And so that if someone does not do this, he is a sleepyhead.)

To which Wittgenstein retorts, in the voice of a complacent calculator:

> I can't imagine a contradiction in my calculus. – You have indeed shewn me a contradiction in another, but not in this one. In this there is none, nor can I see the possibility of one. ... If my conception of the calculus should sometime alter; if its aspect should alter because of some context that I cannot see now – then we'll talk some more about it. (1956, p. 100)

Replacing his "calculus" with our "intended domain of application," Wittgenstein's query becomes "When should the considerations arising within an outside domain D^* be regarded as a reasonable challenge to the reasoning principles we utilize within the domain D?" I do not believe that this question possesses an entirely obvious answer.

3 Logicist Reconstruction

> Numbers are free creations of the human mind; they serve as a means of apprehending more easily and more sharply the difference of things. It is only through the purely logical process of building up the science of numbers and by thus acquiring the continuous number-domain that we are prepared accurately to investigate our notions of space and time by bringing them into relation with this number- domain created in our mind. If we scrutinize closely what is done in counting an aggregate or number of things, we are led to consider the ability of the mind to relate things to things, to let a thing correspond to a thing, or to represent a thing by a thing, an ability without which no thinking is possible. Upon this unique and therefore absolutely indispensable foundation, as I have already affirmed in an announcement of this paper, must, in my judgment, the whole science of numbers be established.
>
> Richard Dedekind (1888, p. 32)

The old-fashioned "persistence of form" justifications of domain extension left nineteenth-century mathematicians in a methodological quandary. Obviously, human thought is greatly advanced by the changes in setting we have documented, but their justificatory methodology appears loose and prone to potential flubs of a Dirichlet principle class. How can mathematics retain its traditional claims to irreproachable certainty if its doctrines hinge upon the mystical whimsies of visionaries such as Jakob Steiner? A popular counterproposal of the time (which assumed several distinct forms) attempted to evade the unregimented vagueness of the "persistence of form" appeals by recasting the extension elements as *logical devices* methodologically legitimated by the primary field from which they spring (viz. from traditional Euclidean geometry in the imaginary points case).

I again remind the reader that I employ the term "logic" in the sense of the "logic" tutors of the time, rather the contents of a modern course in first-order logic. In particular, such primers paid relatively little attention to deductive rules such as *modus ponens* and concentrated instead upon developmental issues that we would now consign to the methodology of science and the psychology of human invention. "Logic" was regarded as laying down the methodological tactics required to recast a loose set of piecemeal discoveries into a properly articulated "science." Dedekind's invocation of "purely logical processes" in our opening motto should be understood as invoking this wider range of considerations. In particular, our modern presumption that "logical reasoning" should remain noncreative and ontologically nonampliative with respect to its base field was entirely alien to the older conceptions I have in mind. It was instead presumed that a proper "logic" should ratify the transformations that allow us to introduce abstract notions such as Motherhood and Patriotism into discourses that previously had lacked these communicative conveniences.

The logic primers of the time commonly called these maneuvers "principles of abstraction" (which they further characterized as "laws of thought"). In John Locke's classic "empiricist" formulation, this takes the form of a rule that generates a new particular (the Abstract Idea) from a collection of particular notions:

> [Children] frame an Idea, which they find those many Particulars do partake in; and to that they give, with others, the name Man, for Example. And thus they come to have a general Name, and a general Idea. Wherein they make nothing new, but only leave out of the complex Idea they had of Peter and James, Mary and Jane, that which is peculiar to each, and retain only what is common to them all. (1979, p. 186)

This process can be mathematized into the modern technique of constructing an array of equivalence classes $\{x|Rxa\}$ by starting with an equivalence relationship R and a base domain **D**. Thus we abstract the rational numbers from the fractions a/b by placing a/b and c/d in the same class if ad = bc. Indeed, a number of various mathematical writers have explicitly equated equivalence class formation with traditional empiricist abstraction. In our times, the technique has become so commonplace that a modern author will reflexively employ the tactic without any deeper consideration of its methodological rationale.

So conceived, equivalence class abstraction represents a methodological transition from particularized *objects* (the fractions) to a new *object* of a more abstract character, the rational number. However, many nineteenth-century authors dismissed this object-to-object abstractionist transition as incoherent: the relationship R must already lie within our conceptual arsenal if we are going to be able to collect objects together as behaving similarly with respect to R. And this line of thought leads to the recognition that certain recipes appeal to evaluative *relationships* as a substitute for an erstwhile guiding object. For example, although we generally rely upon concrete landmarks in providing driving instructions ("Take a 90° turn at the corner where Cazette's jewelry store sits"), a suitable evaluative instruction might serve the same directive purpose ("Turn in the direction common to all of the lines you see in the region".) We know how to evaluate whether two straight lines run parallel to one another or not, and on this basis, the rules of "logic" allow us to frame the abstract notion of a line's direction. This sort of mental process should be regarded as identical in kind to the manner in which we generate the abstract notion of Motherhood from the property of "being a mother." Although we generally employ equivalence class sets for these abstractive purposes nowadays, in the nineteenth century "the direction of line 1" was regarded as simply a primitive sort of "evaluator-object" legitimated by the constructive methodologies of logical thought.[7]

The realization that suitable collections of evaluative traits can behave among themselves like regular "objects" becomes an important structural theme within mathematical thought during the second half of the nineteenth century and continues vigorously through our times as well. I will provide a few remarks on these motivational issues shortly.

The first logicist attempt (that I know of) to avoid appeal to Poncelet's "permanence of relationships" adopts exactly this abstractive tactic for it contends that evaluator-object "directions" can ably play the role of the additional "points at infinity" wanted within extended geometry. In doing so, legitimate forms of logical construction can dislodge misty appeals to the "permanence of relations." The author of this startling proposal (which was subsequently applied by others in many fields) was a German geometer named Karl von Staudt, working between 1844 and 1860. Charlotte Angas Scott characterizes his procedures as follows:

> Von Staudt's primary domain is the visible universe; the elements of his geometry, together with the idea of direction, are an intellectual abstraction

---

[7] I am not aware of any explicit employment of equivalence classes before Dedekind's usage in his writings on algebraic numbers.

from the results of observation. He then extends his domain beyond the
visible universe by formal definition; to replace the idea of direction he
introduces a set of "ideal points", and finds that the nature of an ideal point
is the same as that of a common point. ... While the reason, if sufficiently
trained, is convinced, all natural instincts rebel. (1900, p. 163)

In point of fact, earlier writers such as Steiner himself had already empha-
sized the methodological benefits of conceptually regular objects in terms of
their capacities for evaluating the behaviors of other objects in their environ-
ment. What is a regular point p after all, but a means of evaluating which of its
neighboring lines run through p. (This classificatory collection is traditionally
called a "pencil" within affine geometry.) The underlying insight is behavior-
based: do not conceptualize points as merely structureless objects, but attend
to the manners in which they relate to other objects in their vicinity.
Considered with respect to their operative geometrical functionality, von
Staudt realizes that we can employ a line's direction as a collector of line
"pencils" just as effectively as a regular point p. So why not treat such
directions as novel "points" we can add to the regular Euclidean plane? Von
Staudt then shows that his supplementary evaluator-objects obey reasoning
principles that remain very similar to the rules that prevail within traditional
Euclidean geometry (e.g., every pair of evaluator-object "points" will determine a unique evaluator-object "line" and so forth). The only exception to this
resemblance lies in the fact that the inclusion of the supplementary "points at
infinity" smooth over the deductive barriers that had inspired the introduction
of points at infinity in the first place. In doing so, von Staudt replaces the
dodgy postulations of "permanence of relations" with definitions drawn from
"logic."

Von Staudt's approach to the complex points is even more audacious. In our
discussion of the cat face projection, we noted that when a pole point P lies
outside of the circle C, a natural "involution" mapping will nest itself around the
two points where C intersects the projection screen L. We also noted that
Poncelet's methodology posits supplementary imaginary points to serve as the
missing intersection points for the involution mappings that still "persist" even
after P moves inside C. In the same manner as before, von Staudt utilizes the
evaluations captured within involution mappings as surrogates for the desired
"imaginary points belonging to the circle C." J. L. Coolidge describes these
tactics as follows:

> Von Staudt was acutely conscious that the treatment of imaginary elements in
> pure geometry was extremely unsatisfactory. Poncelet's system of ideal
> chords and supplementaries was the only contribution to the subject that
> had any real substance. He set to work to remedy this defect in truly heroic

fashion. Suppose that on a straight line we have an elliptic involution. A point and its mate in the involution trace the straight line in the same sense. "Very well," says von Staudt, "We will define an elliptic point involution and a sense of description as an imaginary point. The same involution with the contrary sense shall be defined as the conjugate imaginary point." . . . These definitions of von Staudt are certainly revolutionary. It was a bold step to define as an imaginary point something that is made up of an infinite number of real points. Von Staudt could not foresee the analogy to Dedekind's definition of an irrational number as a split in the real number system. What he did do was to show by the most careful reasoning that the new elements thus introduced obeyed just the laws of the old ones. Two of his points determine one of his lines which lies completely in any one of his planes through the two points, etc.[8] (1934, p. 217)

Prima facie, such an approach represents a considerable gain in rigor over the hazy "persistence" of earlier years because the truth-value of every proposition in extended geometry can be fully traced to a regular Euclidean grounding (albeit in a rather weird way). However, the motivations for doing so still trace to the utilities gained through "permanence of relationships" considerations, because the realms of geometry could be likewise inflated in entirely useless ways (e.g., by invoking irrelevant varieties of evaluator-objects such as Motherhood). We will return to this point later.

As I mentioned, the notion that a restricted range of evaluative properties, if collected together in a "space" of their own, can act algebraically as if they comprised a nice set of "objects" in their own right became an important organizational theme within mathematics as a whole.

A nice exemplar of such thinking is found in the modern approach to the vector calculus, in which velocity vectors are regarded as arrow-like objects $\mathbf{v}$ derived from an underlying manifold and collected together into a so-called tangent space $\mathbf{D}$. Dual to $\mathbf{D}$ are the linear functionals (the so-called one-forms) dN that "evaluate" the vectors by mapping them to numerical values (e.g., $dN(\mathbf{v}) = w$, where "w" might represent the work required to move a unit mass along $\mathbf{v}$ within the potential field represented by N). But this special collection of evaluative functionals comprises a "dual space" $\mathbf{D}^*$ in its own right, obeying similar rules for internal multiplication and addition. "Dual spaces" of this character are omnipresent in modern functional analysis, although allied exemplars were already well-known in the 1860s. This is why (for want of a better term) I have called the

---

[8] Coolidge (1934) anachronistically identifies von Staudt's "imaginary points" with equivalence classes rather than primitive *evaluator-objects* in the fashion of a "direction." In fact, von Staudt doesn't clarify his "logical" procedures explicitly.

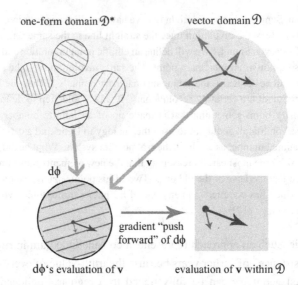

one-form domain $\mathcal{D}^*$ vector domain $\mathcal{D}$

dϕ

v

gradient "push
forward" of dϕ

dϕ's evaluation of **v**   evaluation of **v** within $\mathcal{D}$

Figure 8. Push forward from **D*** to **D**

ingredients within **D*** "evaluator-objects" – to emphasize their hybridized melding of "objects" and "evaluative properties."[9]

In exceptional circumstances (when a "metric" is available within **D**), the evaluators dN within **D*** will possess natural representatives inside **D** itself, called the "push-forwards" of the one-forms dN.[10] This special availability muddied the development of the multivariate calculus for a considerable span of time, but modern approaches to the calculus generally favor a "keep your **D/D*** spaces conceptually distinguished" approach (Fortney 2018). The notion that these special evaluator-object "spaces" should be treated as *standalone universes unto themselves* is central to Dedekind's thinking about "mathematical ontology" in a manner that fails to hold for Gottlob Frege, Bertrand Russell, and W. V. Quine. We will return to these themes in the next section.

Upon reflection, it is clear that we tacitly utilize comparable "spaces" of evaluator-objects within our everyday lives. (Charles Peirce called them "types.") Suppose that a good recipe for designing an English garden involves an intermediate planning stage in which a removable gnome statuette serves as a colorific focus, around which subsequent floral ingredients can are harmoniously arranged. We may consult a catalog for these purposes, whose illustrations provide the "types" of the terra cotta "tokens" we might utilize in our planning. But, as we have already observed, an unrealized gnome evaluation

---

[9] The very term "duality" connotes the fact that these evaluative roles can be reversed, and the vectors viewed as the "properties" that evaluate the one-forms. I believe that Frege's celebrated "context principle" is best understood in this light (Wilson 2010).

[10] In old-fashioned treatments of vector calculus, this pushed-forward one-form is called "grad N."

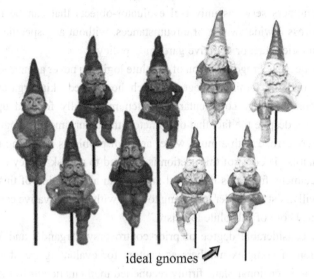

ideal gnomes

Figure 9. Gnome evaluators

can serve our planning purposes just as ably as an actual gnome, given that we did not plan to keep them within our finalized garden.

Indeed, the same garden recipe considerations may prompt us to enlarge the catalog's collection of prototypes with "extension gnomes" (e.g., including, say, an idealized left-handed gnome correspondent to every right-handed specimen found in the catalog). Such an enlarged **D*** collection offers a "better living space" for our aesthetic deliberations.

When nineteenth-century authors write of "logic," they frequently have in mind amplificatory processes of this "fatten up an evaluative space" character (which Dedekind describes as "free creations of the human mind"). As such, the tactic should not be identified solely with the "form an equivalent class based upon the relation R" policies that are most widely employed today, for this approach conceals the radical aspects of von Staudt's approach to the extension element problem. Treatments of the sorts so far surveyed might be dubbed "relative logicisms" because their logically supported evaluator-objects remain tethered to some specific realm of preexistent fact, such as Euclidean geometry or Newtonian mechanical behavior (in the case of the "dual numbers," which we will discuss in a moment). Today philosophers are more conversant with the "absolute logicisms" defended by Dedekind, Frege, Russell, and so forth, in which the familiar number systems (integers, rationals, real and complex numbers) are claimed to represent "purely logical objects" independent of any background applicational basis. This "absolutism" reflects the consideration that the numbers serve as universal evaluator-objects that can be fruitfully utilized across a wide swatch of circumstances, without any specific reliance upon geometrical fact or effective gardening policy.

Largely because Frege's version of absolute logicism never advanced significantly beyond the familiar integers (which he dubbed "kindergarten arithmetic"), contemporary commentary is often myopically focused upon this specific case, despite the fact that our nineteenth-century methodologists were generally concerned with a much wider array of problems involving number-like evaluators. The rest of this section is devoted to a brisk survey of several exotic specimens from this numerical zoo, for an appreciation of this greater expanse will assist the reader in coming to grips with the innovative concerns at the historical core of "absolute logicism."

After a considerable degree of prior controversy, Argand's and Wessel's visualization of complex numbers as vectors for evaluating transformations upon a two-dimensional plane firmly reconciled mathematicians to their legitimacy as certifiable "mathematical entities."

Such numbers provide a convenient algebra for computing the resultants of a sequence of planar rotations and dilations, such as the successive movements of the 2D mechanical assembly illustrated on the left side of Figure 10 (viz., first rotate the A joint through an angle ω_1, then twist joint B by ω_2, and finally slide part C towards D by a length L). Can we concoct a similar algebra of extended "numbers" to perform similar computational duties for the 3D assembly on the right, where we are now allowed to rotate its parts out of the plane of the paper? Framing an algebra that can accomplish this task efficiently (which remains important within modern robotics) requires "numbers" of an unexpectedly

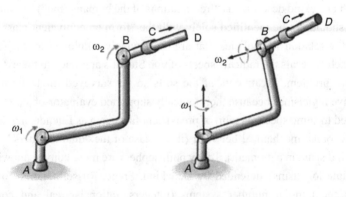

2D composite motion   3D composite motion

Figure 10. Successive machine movements

exotic character (e.g., Hamilton's quaternions or E. Study's dual-numbers).[11] Projects of this kind point to the conceptual necessity of linking any newly extended notion of "number" to a firm basis within a generating set of suitable evaluative standards. If such an algebra can be produced, it automatically supplies the "sharper apprehension" that Dedekind had in mind in our opening epigram: "Numbers are free creations of the human mind; they serve as a means of apprehending more easily and more sharply the difference of things." Such concerns immediately explain why the articulation of a suitable set of number-like objects requires careful scrutiny of the behavioral evaluations they facilitate.

There is a second epicenter of nineteenth-century concern that lies particularly close to Dedekind's heart. Its circumstances nicely illuminate his "structural" reasons for insisting that novel forms of evaluator-object evaluators ought to be collected together into standalone universes that remain algebraically closed unto themselves. Here is a brief sketch of the motivational background. In struggling with problems related to Fermat's last theorem, Ernst Kummer became interested in certain smallish collections of complex numbers in which the property of unique factorization breaks down. Avoiding unnecessary details, here's a simpler situation that illustrates the basic phenomena at issue.[12] Suppose that we are interested in the restricted domain σ of regular integers of the form $4n + 1$ (viz., 1, 5, 9, 13, ...), which remain closed under ordinary multiplication. When we attempt to prove significant facts about σ, we encounter a significant roadblock: the elements of σ do not break uniquely into prime factors (viz., 10857 equals both 141×77 and 21×517). Intuitively, 3 represents a factor common to 21 and 141, but it is not of the form $4n + 1$ and hence does not belong to the domain σ. A mathematician's inferential control over the regular integers relies heavily upon the so-called Euclidean algorithm that uniquely factors integers into their prime factors. If this reasoning tool fails, we automatically lose most of our effective strategies for determining how a domain like σ behaves. Following "persistence of relations" considerations similar to those that Poncelet cited, Kummer removed these reasoning obstructions by extending σ to a richer domain σ^* that includes the desired prime factors (without reducing σ^* to a simple copy of the $4n + 1$ subdomain of integers). Dedekind sought to generalize this tactic for wider purposes and praised Kummer's great discovery as follows:

---

[11] Yang (1974). Frege may have been originally searching for evaluator-objects of this character (Frege 1874).

[12] This example derives from (Hancock 1928). Kummer and Dedekind were primarily concerned with various restricted classes of complex numbers, but Hancock's example is easier to explain.

But the more hopeless one feels about the prospects of later research on such numerical domains, the more one has to admire the steadfast efforts of Kummer, which were finally rewarded by a truly great and fruitful discovery. That geometer succeeded in resolving all the apparent irregularities in the laws [of his target domain σ]. By considering the indecomposable numbers which lack the characteristics of true primes to be products of ideal prime factors whose effect is only apparent when they are combined together, he obtained the surprising result that the laws of divisibility in the numerical domains studied by him were now in complete agreement with those that govern the domain of rational integers. (1877, pp. 56–57)

The rationale that Kummer supplies for this methodological gambit is quite revealing, for he claims that he was merely following the same methodology that allows a chemist to postulate a missing chemical element (fluorine), even though no such ingredient had been isolated up to that time.

Multiplication for [my subvarieties of] complex numbers correspond to the chemical compound; to the chemical elements or atomic weights correspond the prime factors, and the chemical formulae for the decomposition of bodies are precisely the same as the formulae for the factorization of numbers. And the ideal numbers of our theory appear also in chemistry as hypothetical radicals which have not as yet been isolated, but have their reality, as do the ideal numbers in their combination. For example, fluorine, which has not yet been isolated and is still counted as one of the elements, is an analog of an ideal prime factor. Idealism in chemistry however is essentially different from the idealism of complex (algebraic) numbers in that chemical ideal materials when combined with actual materials produce actual compounds: but this is not the case with ideal numbers. Furthermore, in chemistry the materials making up an unknown dissolved body may be tested by means of reagents, which produce precipitates, from which the presence of the various materials may be recognized. Observe that the multiplication of prime ideal numbers produce a rational prime integer, so that the reagents of chemistry are the analogues of the prime ideals, these prime ideals being exactly the same as the insoluble precipitate which, after the application of the reagent, settles.

  Also, the conception of equivalence is practically the same in chemistry as in the theory of algebraic numbers. For, as in chemistry, two weights are called equivalent if they can mutually replace each other either for the purpose of neutralization or to bring about the appearance of isomorphism. Similarly, two ideal numbers are equivalent if each of them can make a real rational number out of the same ideal number.

  These analogies which are set forth here are not to be considered as mere play things of the mind, in that chemistry as well as that portion of the number theory which is here treated have the same basic concept, namely, that of combination even though within different spheres of being. And from this it follows that those things which are related to this [principle of combination] and the given concepts which necessarily follow with it must be found in both [fields] by similar methods: The chemistry on the one hand of natural materials and on the other hand the chemistry of ideal numbers may both be regarded as the realizations of the concept of combination and of the conceptspheres dependent upon this [principle of combination]. The former to be regarded as a physical which is bound with the accidents of external existence and therefore richer, the latter a mathematical, which in its inner necessity is perfectly pure, but therefore also poorer than the former. (1847, p. 356)

Indeed, in this historical period, it was sometimes opined that some chemical "elements" might permanently resist isolation, in the manner that the component quarks within a proton refuse to render themselves individually apparent.

Observe that Kummer suggests that his mathematically extended σ* belongs to a "poorer realm of being" than ordinary chemical compounds – a theme to which we will later return.

Obviously, these loose empiricist-like appeals raise the same specter of untrustworthy grounding that had plagued the projective geometers and the appeals to the Dirichlet principle. In responsive remedy, Dedekind draws a conceptual linkage between the extended reasoning domain σ* and a suitable *evaluative relation*. "What property," he asks, "do the 4 n+1 values 9, 21, 33, . . . share?" *Answer*: They all satisfy a common "divisibility test" that we would characterize as being neatly divisible by 3 from an external perspective, despite the fact that 3 does not belong to σ (so our evaluative relation is "x agrees with y under the specified divisibility test"). Dedekind calls the resulting equivalence classes "ideals" and collects them together into an amplified algebraic domain σ*.[13] Dedekind explains these procedures as follows:

Kummer did not define ideal numbers themselves, but only the divisibility of these numbers. If a number a has a certain property *A*, to the effect that a satisfies one or more congruences, he says that a is divisible by an ideal number corresponding to the property *A*. While this introduction of new numbers is entirely legitimate, it is nevertheless to be feared at first that the language which speaks of ideal numbers being determined by their products, presumably in analogy with the theory of rational numbers, may lead to hasty conclusions and incomplete proofs. And in fact this danger is not always completely avoided. On the other hand, a precise definition covering all the ideal numbers that may be introduced in a particular numerical domain o, and at the same time a general definition of their multiplication, seems all the more necessary since the ideal numbers do not actually exist in the numerical domain o. To satisfy these demands it will be necessary and sufficient to establish once and for all the common characteristic of the properties *A, B, C,* . . . that serve to introduce the ideal numbers, and then to indicate how one

---

[13] The members of σ possess natural embeddings into σ* as "prime ideals," but we need not pursue these arrangements further here.

can derive, from properties *A, B* corresponding to particular ideal numbers, the property *C* corresponding to their product.

This problem is essentially simplified by the following considerations. Since a characteristic property *A* serves to define, not an ideal number itself, but only the divisibility of the numbers in o by the ideal number, one is naturally led to consider the set α of all numbers a of the domain o which are divisible by a particular ideal number. I now call such a system an ideal for short, so that for each particular ideal number there corresponds a particular ideal a. Now if, conversely, the property *A* of divisibility of a number a by an ideal number is equivalent to the membership of a in the corresponding ideal **a,** one can consider, in place of the properties *A, B, C,* ... defining the ideal numbers, the corresponding ideals **a, b, c,** ..., in order to establish their common and exclusive character. Bearing in mind that these ideal numbers are introduced with no other goal than restoring the laws of divisibility in the numerical domain o to complete conformity with the theory of rational numbers, it is evidently necessary that the numbers actually existing in o, and which are always present as factors of composite numbers, be regarded as a special case of ideal numbers. (1877, pp. 57–58)

In this last remark, Dedekind observes that (in our 4 n + 1 illustration) the original σ-domain object 5 has a natural representative within σ* in the form of the "prime ideal" {5, 10, 15, ...}, whereas {9, 21, 33, ...} represents something novel that only appears within σ*. (This supplementary "ideal" performs the factorization chores of our missing 3 factor.)

In contrast to an author like Frege, Dedekind articulates two characteristic themes with respect to this construction technique. (i) There is nothing sacrosanct or essential about the {5, 10, 15, ...} identification based upon σ. As a closed-unto-itself dominion, σ* constitutes an algebraic structure that reappears across mathematics in many guises dissociated with the base domain σ. Dedekind views his equivalence class construction simply as a convenient ladder that utilizes the firm facts within σ as a means of verifying that the facts internally pertinent to σ* have been adequately specified (so that we need not worry about "6/0" woes arising when we utilize σ*'s algebra). However, σ*'s internal computational soundness could have been equally certified by a range of alternative ladder-like constructions that rely neither upon σ nor Dedekind's ideals. (ii) The motivational insight that σ* represents a useful extension domain in which tricky questions about numerical behavior can be more easily unraveled belongs entirely to Kummer, for Dedekind's after-the-fact constructions do not explicate the diagnostic utilities that render Kummer's enlargement so fruitful mathematically. Borrowing terminology from Section 2, Dedekind believes that he has merely converted Kummer's creative "Art" into a more regular "Science," potentially applicable within other fields other than Kummer's numerical interests. He writes:

A special definition is therefore needed in order to admit negative multipliers as well, and thereby to liberate the operation from the initial constraint [of only applying within o]; but such a definition involves a priori complete arbitrariness, and it would only later be decided whether then this arbitrarily chosen definition would bring any real use to arithmetic; and even if the definition succeeded, one could only call it a lucky guess, a happy coincidence – the sort of thing a scientific method ought to avoid. So let us instead apply our general principle [of persistence of theorems]. We must investigate which laws govern the product if the multiplier undergoes in succession the same general alterations which led to the creation of the sequence of negative integers out of the sequence of positive integers. (1854, p. 758)

In writing of a "persistence of theorems," Dedekind self-consciously echoes the manner in which earlier conceptions of extension elements had been regarded as inspired by the "permanence of relations" or the "persistence of form." He recommends that such considerations should be rendered more "rigorous" by attending to the problem of certifying that every claim expressible within the vocabulary of σ* has been assigned a crisp truth-value. He accomplishes this specific task by appending suitable evaluator-objects to σ, although he recognizes that other "constructive ladders" may serve the same purpose equally well. These duties of rigorization are vital to mathematics' continued good health, but they do not in themselves explain why mathematicians should want to extend σ to σ* in the first place. (Why not add in Motherhood instead?) To address this second concern, Dedekind insists that we must bear in mind the practical considerations prompted "geniuses" such as Poncelet and Kummer to recognize that certain special forms of extension element supplement comprise natural "organic outgrowths" of ingredients that had already certified their centrality within mathematical culture. This theme becomes central within our Section 8.

4 Set Theoretic Ladders

If space has at all a real existence it is not necessary for it to be continuous; many of its properties would remain the same even were it discontinuous. And if we knew for certain that space was discontinuous there would be nothing to prevent us, in case we so desired, from filling up its gaps, in thought, and thus making it continuous; this filling up would consist in a creation of new point-individuals and would have to be effected in accordance with the above principle.

Richard Dedekind (1872, p. 12)

A distinctive feature of Dedekind's point of view lies in the fact that he does not regard his own "scientific" improvements as necessarily capturing the considerations that first suggest that an older domain of study might profit from an

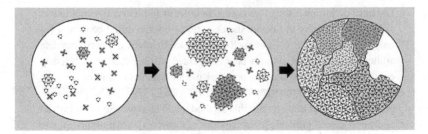

Figure 11. Crystallization from a melt

"organic" recasting with respect to its "setting." Nor does he demand that the entirety of mathematics must stem from a single motivational core. Instead, its ever-enlarging topics should gradually consolidate themselves in the manner of a crystalline conglomerate emerging from a mineral melt. Beginning in insightful acts of inventive "art," nucleating centers form within the ambient lava, which subsequently enlarge into thicker blocks through the systematizing activities natural to a "science."

Eventually these various blocks begin to jam against one another, mutually inhibiting their unfettered outgrowth. A methodologist of a "scientific" disposition might assist the stability of the final composite by smoothing off their interfacial joins, but the central motivations that determine its overall structural organization continue to trace to the insights of the Cardanos, Eulers, and Kummers, who provided the original nucleating seeds. This acceptance of inherent motivational diversity will become central in our concluding section.

For these reasons, we should not presume that the set theoretic "ladders" that Dedekind frequently employs in his efforts at "scientific" consolidation necessarily reflect mathematics' motivational "ontology" in any significant manner. In an often-cited letter to his coauthor Heinrich Weber, Dedekind warns against such a hasty identification:

> But if one were to take your route – and I would strongly urge that it be explored once to the end – then I would advise that by [cardinal number] one understand not the class itself (the system of all finite systems that are similar to each other) but something new (corresponding to this class) which the mind creates. We are a divine race and undoubtedly possess creative power, not merely in material things (railways, telegraphs) but especially in things of the mind. This is precisely the same question that you raise at the end of your letter in connection with my theory of irrationals, where you say that the irrational number is nothing other than the cut itself, while I prefer to create something new (different from the cut) that corresponds to the cut and of which I say that it brings forth, creates the cut. We have the right to ascribe such a creative power to ourselves; and moreover, because of the similarity of

all numbers, it is more expedient to proceed in this way. The rational numbers also produce cuts, but I would certainly not call the rational number identical with the cut it produces; and after the introduction of the irrational numbers one will often speak of cut-phenomena with such expressions, and inscribe to them such attributes, as would sound in the highest degree peculiar were they to be applied to the numbers themselves. Something quite similar bolds for the definition of cardinal number as a class; one will say many things about the class (e.g. that it is a system of infinitely many elements, namely, of all similar systems) that one would apply to the number only with the greatest reluctance; does anybody think, or won't he gladly forget, that the number four is a system of infinitely many elements? (But that the number four is the child of the number three and the mother of the number five is something that nobody will forget.) For the same reason, I always considered Kummer's creation of the ideal numbers to be thoroughly justified, if only it were rigorously carried out. Whether in addition the language of symbols suffices to designate uniquely each individual that is to be created, is not important; it always suffices to designate the individuals that appear in any (limited) investigation. (1888a, pp. 834–35)

In writing of "cuts," Dedekind here refers to his celebrated technique of building up the real numbers from equivalence classes of fractions.

By this time verification policies that we might now regard as "crypto-set theoretic" had already proved central within many parts of mathematics, due to an enlarging appreciation of the difficulties residing in naïve appeals to "limits." These troubles stem from the fact that the mathematical characteristics of functions constructed through limiting operations frequently differ significantly from the base functions that lead up them. The canonical illustration is due to Fourier: a set of ordinary sine functions (which are "smooth" in the sense of possessing well-defined derivatives everywhere) can unexpectedly produce derivative-less "jumps" in their limiting product. More generally, our intuitive expectations with respect to limits are notoriously unreliable, and mathematicians learned to never presume without proof that a characteristic ϕ maintained throughout the limiting sequence $f_0, f_1, f_2, \ldots$ will continue to hold in the limit f_∞. Allied considerations warn that we should become wary of the misleading suggestions of functional expressions such as a purported integration "$\int f_i \, dx$". The little pieces that one intuitively "adds together" may not approach the smooth pointwise result one anticipates in a regular, pointwise manner but may oscillate wildly around it without a final assignable value.[14]

---

[14] Mathematicians know that they must carefully verify that syntactic expressions such as "$\int f_i \, dx$" qualify as truly denoting. Indeed, Friedrich Prym constructed an early counterexample to the Dirichlet Principle that fails in an oscillatory manner (similar phenomena create difficulties in bringing mechanical interiors in accord with their surrounding boundaries, as we shall note in Section 6). Russell's theory of descriptions was originally devised as a means of addressing these difficulties.

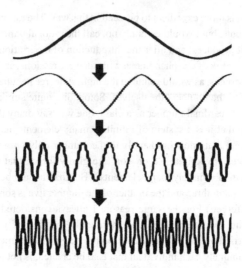

Figure 12. Oscillating limit

Methodologists like Dedekind realized that these difficulties frequently rely upon assumptions that we now call "completions of a space." If we articulate criteria for a "space" that contains the preliminary sequence $f_0, f_1, f_2, \ldots$, will all of their presumptive limits f_∞ satisfy these criteria, or will they instead represent "holes" that require after-the-fact filling in? In a geometrical context, starting with a unit upon a base line L, we can progressively reach to a large number of points along L via a set of natural constructions that von Staudt called "the calculus of throws." But will these constructive tactics actually cover every real-valued position along L, or will various "holes" be left unreached? In fact, the latter possibility holds, and a supplementary "completeness assumption" is required to close the gaps. Although various authors claimed that these filled-in ingredients derive their methodological support from "geometrical intuition," Dedekind draws an entirely opposite conclusion, for he contends that the supportive rationale for these completions instead trace to our capacities for "logical" supplementation, as he expresses in the opening motto to this section. It is entirely the "free creativity" of the methodologist that recommends the transition from intuition-based geometry to its more convenient "completion space." Just as we witnessed with his algebraic "ideals," the set-theoretic details of Dedekind's celebrated "cut constructions" merely supply one of the various ladders that one might employ to verify that our "completion" possesses adequately defined truth-values everywhere. As soon as this transitional task is accomplished, we may throw away the particular ladder utilized for this purpose. Set theory offers an extremely useful toolkit for underwriting these extensions, but it does not directly explain why we find them useful. Assigning the activities of "completion" to our general "logical" faculties prevents us from falling into presumptive error with respect to their "intuitive" support.

To be sure, set theoretic deliberation provides the court of final appeal that modern mathematicians utilize in addressing the existence worries that frequently plague the employments of integral and differential equations, but this important utility does not otherwise confer any "ontological" centrality upon set theoretic thinking.

Dedekind's letter to Weber is often cited as a precursor to the "ontic structuralisms" advanced by Paul Benacerraf, Stewart Shapiro, and other modern writers, in which "mathematical structures are well-defined only up to isomorphism." Following this suggestion, mathematics is viewed as an all-embracing "universe of structures," built up in the general manner of Zermelo-Fraenkel set theory, but with their unwanted ε-dependent filigree erased. For the reasons just outlined, I do not detect similar ambitions within Dedekind's own thought, who continually emphasizes that he is largely interested in the "structures" that represent the "organically motivated" enlargements of well-established earlier domains:

> This way of proceeding repeatedly leads us to new number domains, because the previous domain no longer satisfies the demand for the general applicability of the arithmetical operations; one is thereby compelled to create the irrational numbers (with which the concept of limit appears) and finally also the imaginary numbers. These steps forward are so immense that it is difficult to decide which of the many different paths which open before us we should first pursue. But obviously, if one wants to apply to these new classes of numbers the operations of arithmetic as they have hitherto been developed, then repeated extensions of the earlier definitions are requisite, and here, at least with the appearance of imaginary numbers, the chief difficulties of systematic arithmetic begin. However, we may well hope that we will attain to a truly solid edifice of arithmetic if we here too persistently apply the fundamental principle: not to permit ourselves any arbitrariness, but always to allow ourselves to be led on by the discovered laws. As is well known, an unobjectionable theory of the imaginaries (not to mention the numbers newly invented by Hamilton) does not exist – or at any rate it has not yet been published. (1854, p. 757)

Insofar as I can see, modern readers of Dedekind frequently overlook this patchwork-based portrait of mathematical doctrine, leaving them only with a bare "structuralism" supplied by appeals to mysterious appeals to "free creativity" that John Burgess characterizes as "mystical":[15] "To the question

---

[15] I suggest a more sympathetic evaluation of Dedekind's scattered appeals to "genius" and "free creativity" in Section 8.

whether we may still speak of the natural number system, the mystics . . . answer 'Yes, but the structure in question is a very peculiar one'" (2015, p. 132).

A further characteristic of Dedekind's thought is that he concentrates upon subjecting specific domains of intuitive reasoning to "scientifically" certified innovation improvement by filling in their internal algebraic behaviors with useful "logical" extensions, largely leaving aside the further question of where these perfected blocks of reasoning tools might someday find a useful application. But it is the latter form of ratification that ultimately determines whether these improvements in reasoning technique actually deserve a permanent position within the evolving arsenal of profitable mathematics. Hermann Weyl articulates an allied modesty with respect to some of his own work:

> But I do not want to conceal from you the growing feeling among mathematicians that the fruitfulness of the abstracting method is close to exhaustion. It is a fact that beautiful general concepts do not drop out of the sky. The truth is that, to begin with, there are definite concrete problems, with all their undivided complexity, and these must be conquered by individuals relying on brute force. Only then come the axiomatizers and conclude that instead of straining to break in the door and bloodying one's hands one should have first constructed a magic key of such and such shape and then the door would have opened quietly, as if by itself. But they can construct the key only because the successful breakthrough enables them to study the lock front and back, from the outside and from the inside. Before we can generalize, formalize and axiomatize there must be mathematical substance. (2002, p. 161)

We can sharply articulate an important divergence between Dedekind and Frege on these issues if we return to the dualized one-form/vector domains $\mathbf{D}^*$ and $\mathbf{D}$ discussed in the previous section. We observed that $\mathbf{D}^*$ is comprised of the linear functionals that "evaluate" the features of the vectorial objects $\mathbf{v}$ residing within $\mathbf{D}$. In specialized forms of these $\mathbf{D}^*/\mathbf{D}$ pairings (where a metric is available), every $d\phi$ within $\mathbf{D}^*$ will possess a natural $\mathbf{D}$-domain representative $\mathbf{v}$ (traditionally called grad(ϕ)) that can duplicate $d\phi$'s $\mathbf{D}^*$-upon-$\mathbf{D}$ actions while remaining entirely inside $\mathbf{D}$. (This $\mathbf{v}$ surrogate is often called the "pushforward" of $d\phi$.) Traditional treatments of the calculus usually present the one-form evaluations constitutive of $\mathbf{D}^*$ in this "pushed-forward" guise, but modern texts generally reject this amalgamation of domains as a mistake, because the $\mathbf{D}^*/\mathbf{D}$ pairings must be kept distinguished within more general forms of "Banach space" arrangement. As I understand his intentions, Frege's approach to numbers represents an attempt to begin within the specialized subdomain $\mathbf{D}^*$ of concept functions that evaluate the characteristics of regular concepts ψ such as "being a human being" (i.e., they evaluate ψ for cardinality). Frege presumes that he should then push these logical evaluators forward into the wider ontological realm of "objects in general." In doing so, he falls victim to the celebrated Russell paradox. Dedekind, in contrast, adheres to the modern recommendation to keep ones scientifically improved reasoning domains segregated from one another, without evident concern for establishing a unified ontological realm in Frege's manner. Mathematics' repertory of increasingly abstracted structures (such as his factorization domains σ*) should generally "grow organically" from the seeds of previously established solidified domains of mathematical interest. But Dedekind never pretends to envision where mathematics may eventually wind up in its continued dogged pursuit of useful reasoning tools.

5 If-Thenism

> In my opinion, a mathematician, in so far as he is a mathematician, need not preoccupy himself with philosophy – an opinion, moreover, which has been expressed by many philosophers.
>
> Henri Lebesque (Bell 1937, p. xvii)

We noted that at the end of the nineteenth century, mathematician Charlotte Angus Scott expressed admiration for von Staudt's rigorizations and regarded them as essential to the conceptual integrity of complexified synthetic geometry. She also realized that his works were virtually unreadable, in a manner "against which all natural instincts rebel." This was not an infrequent sentiment; Hermann Hankel had satirized his labors as follows:

> The work of von Staudt, classical in its originality, is one of those attempts to force the manifoldness of nature with its thousand threads running hither and thither into an abstract scheme and an artificial system: an attempt such as is only possible in our Fatherland, a country of strict scholastic method, and, we may add, of scientific pedantry. The French certainly do as much in the exact sciences as the Germans, but they take the instruments wherever they find them, do not sacrifice intuitive evidence to a love of system nor the facility of method to its purity. In the quiet town of Erlangen, von Staudt might well develop for himself in seclusion his scientific system, which he would only now and then explain at his desk to one or two pupils. In Paris, in vivid intercourse with colleagues and numerous pupils, the elaboration of the system would have been impossible. (1875, p. 669)

But continued interest in von Staudt's project vanished virtually overnight after Hilbert's celebrated resuscitation of Euclidean-style axiomatization emerged in 1899, for it appears to justify a simple methodology that Bertrand Russell later dubbed as *if-thenism*: "a mathematician is free to explore any self-consistent axiom scheme of her choice." (These axioms provide the "if" part of Russell's

scheme, and their resulting theorems constitute their subsequent "then" consequences.) From this perspective, the projective geometers can directly articulate
their own parochial axioms for whatever enlarged realms they desire,[16] and older
neurotic concerns with respect to their justificatory status can be swiftly set aside
in favor of inquiries into how the interior models of these otherwise self-standing
schemes relate to one another. The pedagogical relief that subsequent authors
derive from this axiomatic liberation from justificatory logicisms of a von Staudt-
style character is palpable in later textbooks:

> Questions probably arise immediately in your mind: "What right have we
> arbitrarily to add points to space? And, even if we may be permitted to do so,
> must we not differentiate between the "ordinary points," which we see as dots
> on the paper, and the "ideal points," which are constructs of our
> imagination?" ... [Geometry] deals with planes, straight lines, and points.
> These are entities about which certain assumptions (axioms, postulates) are
> made but which are otherwise undefined. Hence, we can give the name
> "point" to anything which conforms to our postulates about points and
> which suits our convenience to designate as "point". In short, all our geom-
> etry (indeed, all our mathematics) deals with constructs of our imagination;
> and an "ideal" point is on precisely the same footing, from this point of view,
> as an "ordinary" point. (Rosenbaum 1963, p. 16)

Implicit within these appeals is a widely shared hope that by characterizing
their investigative procedures in sufficient detail, working scientists can wall off
their investigations from inconclusive "philosophical" considerations of how
their words relate to the external world or to wider considerations of human
psychology. Underlying these attitudes is the pragmatic presumption that if
procedures have been specified that can completely resolve the truth-values of
all of the sentences constructible within a fixed grammatical field, then the
relevant propositions have been assigned "meanings" that are entirely "adequate
for science's purposes." Such presumptions lie behind H. G. Forder's remark:

> Our Geometry is an abstract Geometry. The reasoning could be followed by
> a disembodied spirit who had no idea of a physical point; just as a man blind
> from birth could understand the Electromagnetic Theory of Light ...
> Theoretically, figures are unnecessary; they are [solely] needed as a prop to
> human infirmity. (1927, p. 43)

We might observe that similar themes arise with respect to the vocabularies of
physics, with a parallel hope that pointless arguments about "the meaning of
'force'" can be skirted by "implicitly defining" terminologies within an
adequately specified axiomatic housing. The notion that meanings can be

---

[16] This was supplied in (Veblen and Young 1910).

"implicitly defined" within a properly regimented framework served as an attractive flame around which the moths of twentieth-century philosophy constantly fluttered. However, we cannot explore these wider thematic ramifications here.

With respect to real-life mathematical practice, these "official" professions of liberal tolerance are deeply insincere. Few working mathematicians accept the wide permissions of genuine "if-thenism," and X's endeavors within field Y will often become scornfully dismissed on the grounds that "X has not worked in the proper setting." Mathematicians are sometimes surprisingly cruel in how they patronize their colleagues along these illiberal contours. Nor do they truly accept formalist allowances with respect to "truth-value." In Section 7, we will discuss Oliver Heaviside's self-improvised forays into his so-called "operational calculus," which uncovered novel solutions to differential equations in very peculiar ways. As a result, his writings were rejected as "unrigorous" by "mathematicians of the Cambridge or conservatory kind who look the gift-horse in the mouth and shake their heads with a solemn smile" (Heaviside 2003, II, p. 12).

Today it would be universally acknowledged that Heaviside had been correct in most of his innovations, and his Cambridge critics had evaluated their truth-values "within the wrong setting" (which did not submit to axiomatic articulation for nearly sixty years). It qualifies as a classic illustration of what Philip Davis has in mind by "impossibilities being broken by the introduction of structural changes."

Despite these inconsistencies, whenever we press professional mathematicians for their "governing philosophies," they usually offer some variant on ultra-permissive "if-thenism." ("I merely investigate what follows if postulates A are accepted.") As indicated previously, I believe that they do so in an effort to avoid the discomfiting terrain of "foundations of mathematics" dialectic ("Are you a Platonist?"; "Why are nonconstructive forms of proof acceptable?," etc.). I consider this pronounced aversion to "philosophy" to be quite regrettable because the modes of reconsideration that commonly prompt significant "changes in setting" ought to qualify as deeply "philosophical" in the best sense possible, although we will not pursue this theme further here.

"If-thenism" retrenchment allows practitioners to maintain that their labors qualify as "permanent monuments to a priori truth" in the manner outlined in Section 1. Traditional Euclidean geometry can retain a qualified "necessity" as a branch of "if-then" mathematics, despite having lost its erstwhile prestige as the "physical geometry" to be applied to the universe around us. This diminishment in direct relevance has troubled some observers, as James Byrnie Shaw complains:

Geometry appeared to be that branch of applied mathematics which had invincible truth as its character, and which was the ideal toward which all science must strive. Systems of philosophy were founded upon it, and it was the pride of the intellectual world. Yet what a contrast between this height and the modern axiomatic treatment of geometry, in which almost any conceivable set of definitions which are not logically inconsistent, though they may sound absurd, may be used as the starting-point of a game called geometry, whose artificial rules and abstract situations have little to do with human experience apparently, or at most are convenient in the same sense that a meter stick is useful, or equally the king's arm, or the pace he sets. We assume objects A, B, etc., are merely distinguishable from each other; for example, stars and daggers will do. We order them according to certain arbitrary rules. We set down the logical deductions therefrom, and we have a geometry on a postulational basis. Is this the last word and has reality vanished into vacuity and mathematics into a game of solitaire? (1918, p. 32)

Poincaré articulates this tension (which comprises the animating motif of this book) as a dialectical dilemma:

The very possibility of the science of mathematics seems an insoluble contradiction. If this science is deductive only in appearance, whence does it derive that perfect rigor no one dreams of doubting? If, on the contrary, all the propositions it enunciates can be deduced one from another by the rules of formal logic, why is not mathematics reduced to an immense tautology?
  ... [I]t must be conceded that mathematical reasoning has of itself a sort of creative virtue and consequently differs from a syllogism. (1902, p. 31)

A one-sided focus upon the local requirements of "proper scientific development" has led many modern authors to characterize the occasional "mistakes" of pioneers like Euler and Riemann as stemming from the fact that they did not specify the formal setting in which they worked adequately. In *Divergent Series*, G.H. Hardy writes:

This remark is trivial now: it does not occur to a modern mathematician that a collection of mathematical symbols should have a "meaning" until one has been assigned to it by definition. It was not a triviality even to the greatest mathematicians of the eighteenth century. They had not the habit of definition: it was not natural to them to say, in so many words, "by X we mean Y". There are reservations to be made, ... but it is broadly true to say that mathematicians before Cauchy asked not "How shall we define $1-1+1-\ldots$?" but "What is $1-1+1-\ldots$?", and that this habit of mind led them into unnecessary perplexities and controversies which were often really verbal.
  The results of the formal calculations [I offered earlier] are correct whenever they can be checked: thus all of the formulae ... are correct. It is natural to suppose that the other formulae will prove to be correct, and our transformations justifiable, if they are interpreted appropriately. We should

then be able to regard the transformations as shorthand representations of more complex processes justifiable by the ordinary canons of analysis. It is plain that the first steps towards such an interpretation must be some definition, or definitions, of the "sum" of an infinite series, more widely applicable than the classical definition of Cauchy. ...

It is easy now to pick out one cause which aggravated this tendency, and made it harder for the older analysts to take the modern, more "conventional", view. It generally seems that there is only one sum which it is "reasonable" to assign to a divergent series: thus all "natural" calculations with the series ... seem to point to the conclusion that its sum should be taken to be 1/2. We can devise arguments leading to a different value, but it always seems as if, when we use them, we are somehow "not playing the game." (1956, pp. 5–6)

Sometimes these exploratory policies are characterized as a variety of headstrong mania:

The great creators of the infinitesimal calculus – Leibniz and Newton – and the no less famous men who developed it, of whom Euler is the chief representative, were too intoxicated by the mighty stream of learning springing from the newly-discovered sources to feel obligated to criticize fundamentals. (Knopp 1990, p. 1)

And Jean Dieudonné echoes this latter-day triumphalism as follows:

It is certainly incorrect to attribute this attitude (as many authors do) to an unending evolution of a general concept of "rigor" in mathematical arguments, which would be doomed to perpetual change; what history shows us is a sectorial evolution of "rigor." Having come long before "abstract" algebra, the proofs in algebra or number theory have never been challenged; around 1880 the canon of "Weierstrassian rigor" in classical analysis gained wide acceptance among analysts, and has never been modified... It was only after 1910 that uniform standards of what constitutes a correct proof became universally accepted in topology, with Brouwer's work on simplicial approximation and Weyl's treatment of the theory of Riemann surfaces; again, this standard has remained unchanged ever since. (2009, pp. 15–16)

There is no doubt that a warmer regard for "anti-counterexample" rigor has proved a great boon to mathematical research, just as Dedekind's insistence upon a "properly scientific development" reflects a kindred eagerness to avoid unhappy surprises of a Dirichlet Principle character. But even a supreme advocate of the utilities of axiomatic articulation like David Hilbert regarded the purist isolationism articulated by Hardy as excessive:

Surely the first and oldest problems in every branch of mathematics spring from experience and are suggested by the world of external phenomena. Even the rules of calculation with integers must have been discovered in this fashion in a lower stage of human civilization ... and the same is true of the first problems of geometry, ... the theory of curves and the differential and integral calculus, the calculus of variations, the theory of Fourier series and the theory of potential. ... [But in] the further development of a branch of mathematics, the human mind, encouraged by the success of its solutions, becomes conscious of its independence. It evolves from itself alone, often without appreciable influence from without, by means of logical combin-ation, generalization, specialization, by separating and collecting ideas in fortunate ways, new and fruitful problems, and appears then itself as the real questioner. ... In the meantime, while the creative power of pure reason is at work, the outer world again comes into play, forces upon us new questions from actual experience, opens up new branches of mathematics, and while we seek to conquer these new fields of knowledge for the realm of pure thought, we often find the answers to old unsolved problems. ... And it seems to me that the numerous and surprising analogies ... which the mathematician so often perceives in the methods ... of his science, have their origin in this ever-recurring interplay between thought and experience. ... A new problem, especially when it comes from the world of outer experience, is like a young twig, which thrives and bears fruit only when it is grafted carefully and in accordance with strict horticultural rules upon the old stem, the established achievements of our mathematical science. (1902, p. 437)

To be sure, Hilbert also believes that such "outside" proposals should not be regarded as "properly mathematical" until they have been suitably organized into an axiomatic framework, driven by the imperatives of an alleged "universal philosophical necessity of our understanding." (We will return to these "univer-sal imperatives" in Section 8, in connection with similar comments from Dedekind himself.) But Hilbert clearly acknowledges, in a manner that Hardy does not, that a significant portion of developmental thinking must remain *outside* of the narrow confines of the "properly mathematical" as Hilbert conceives them (which may not be able to accommodate the taut demands of axiomatics until its recalibrated subject matter has reached a robust maturity, if it ever happens). He would not wish to dismiss these netherworld exploratory processes as not "part of mathematics" more widely construed.

As is well known, Hardy simultaneously embraced a notoriously other-worldly form of Platonism:

> The mathematician's patterns, like the painter's or the poet's must be beauti-ful; the ideas like the colors or the words, must fit together in a harmonious way. [But the] mathematician ... has no material to work with but ideas, and so his patterns are likely to last longer, since ideas waste less time than words. Beauty is the first test: there is no permanent place in the world for ugly mathematics.[17] (1940, pp. 13–14)

---

[17] Of course, Hardy is not the least bit apologetic here, in any normal sense of the term.

It is this same alleged "beauty" that distinguishes the good "definitions" from the ugly ones. It is through appeal to these misty aesthetics that officially tolerant "if-thenism" advocates wind up policing their subjects in decidedly unliberal ways.

With respect to Hardy's demands for "habits of definition," in many situations, this request cannot be fulfilled readily. Such concerns particularly apply to a large class of important "divergent series expansions" that Hardy does not mention within his survey at all, despite their centrality within most branches of applied mathematics. These are the puzzling "asymptotic expansions" that supply quite accurate values for differential equations if their lowest terms are utilized but fail to converge by emitting utter rubbish once we move on to subsequent terms in the expansion. Despite this peculiar foible of late-term divergence, these formulas are tremendously informative in their utilities. Mathematicians sometimes possess well-behaved convergent expressions for certain physical traits (e.g., Airy's series for the bands in a rainbow), but thousands of their terms must become added together before the partial summations remotely resemble the patterns witnessed in nature. But greater descriptive accuracy can be extracted from three or four initial terms within a suitably devised asymptotic expansion (as long as we stop our summations before the "bad stuff" begins creeping in). Somehow the divergent expansion encodes valuable information about our rainbow in a vastly more accessible manner than its convergent cousins. I do not believe that we fully understand how this "encoding" operates, despite the fact that practical science is everywhere stitched together with liberal applications of these mysterious techniques. If we insisted upon the unavailable "definitional habits" that Hardy demands, the entire fabric would collapse into tatters.

To be sure, we should heed Augustus De Morgan's warnings to the extent we are able: "The student should avoid founding results upon divergent series, as the question of their legitimacy is disputed upon grounds to which no answer commanding anything like general assent has yet been given. But they may be used as a means of discovery" (1849, p. 55). We will examine some of De Morgan's nuanced opinions on such matters in the next section.

As already noted, one of the twentieth century's greatest achievements in clarifying the "organic" interconnections between distinct swatches of the established explanatory pattern was achieved by Laurent Schwartz in his "theory of distributions" of 1950, which provides strikingly novel recastings of well-established differential equation models. If we ignore these wider advantages and attend only to the definitional tweakings he offers, the revisions can invite the complaint: "If you decide to call your dog a 'lion,' petting lions will not seem dangerous." In fact, Schwartz himself reports upon a facetious version of this very criticism:

> Tuckey ... published a long article full of humor on forty mathematical lessons for hunting a lion ... He then went on to define a method for hunting lions using distributions. Anything in the desert is a lion, but in the sense of distributions. For example, a stone is a lion in the sense of distributions. (2000, p. 247)

As we observed, Schwartz's proposals represent an "organic" refashioning of older materials in a manner of which Dedekind would have wholly approved. But if mathematics truly embraced the liberal tolerances offered by "if-thenism," then distinguishing Schwartz's advances from a pack of "call a dog a 'lion'" imposters will seem like a very subjective discrimination.

Before moving on, we should observe in passing that Hilbert's original hope that the internal consistency of any axiomatized subject matter can be established through elementary arithmetical reasoning fell victim to Kurt Gödel's celebrated incompleteness results. In consequence, set theoretic examination continues to provide the central set of "ladders" that mathematicians utilize to resolve the recalcitrant "existence" concerns of differential equations and their equally unruly cousins. I find it hard to discern much of Hardy's "beauty" within these ugly – yet wholly necessary – ratifications.

In any case, in considering a lengthy reasoning process that runs task 1 → task 2 → ... → task n, each component within the routine may have been rigorously checked against potential counterexample in the manner Hardy desires, but the manner in which they have become connected in sequence has not. (We will discuss an example in the next section.) Within the joints interconnecting the components, a large degree of novel reasoning technique can continually amplify our descriptive practices, although we may not detect these innovations if we concentrate too myopically on whether the "certainties" within the component tasks can be established under the umbrella of a suitable set of "habit of definition" axiomatics.

6 Exploratory Mathematics

* Every operation in mathematics that we can invent amounts to asking a question, and this question may or may not have an answer according to circumstances. If we write down the symbols for the answer to the question in any of those cases where there is no answer, and then speak of them as if they meant something, we shall talk nonsense. But this nonsense is not to be thrown away as useless rubbish. We have learned by very long and varied experience that nothing is more valuable than the nonsense which we get in this way; only it is to be recognized as nonsense, and by

means of that recognition made into sense. We turn the nonsense into sense by giving a new meaning to the words or symbols which shall enable the question to have an answer that previously had no answer.

William Kingdon Clifford (1891, pp. 33–34)[18]

One of the disconcerting features of "if-thenism" is that it relegates much of what we would normally regard as significant "mathematical advance" to a hazy dominion of pre-doctrinal exploration. As noted in Section 1, Euler originally mapped out the basic behaviors of complex functions through trial-and-error computational forays into unknown regions offering "*a posteriori* observations which have little chance of being discovered *a priori.*" (Euler 1761, p. 657) The results were often so surprising that he observed that "his science in the person of his pencil seemed to surpass himself in intelligence."[19] In conducting these quasi-experimental searches, Euler was, in fact, canvassing the camouflaged terrain of Riemann surfaces without realizing how strange lay the ground underneath. When Hardy complains that figures like Euler lacked "the habits of definition," he suggests that a modern Euler would have defined his functions properly before he published any of his exploratory results. "To be sure, every novel mathematical proposition," Hardy would have conceded, "requires a lot of preliminary scribbling on coffee house napkins and the like, but the results shouldn't qualify as 'proper mathematics' until a rigorous proof has been articulated in terms of well-articulated premises and definitions. Before that stage, we should only speak of speculative conjectures." Availing himself of this distinction, Hardy can defend the traditional presumption that "proper mathematics" represents a repository of unshakeable truths, relegating Euler's actual labors to the preparatory antechamber.

To counteract this popular position, we will review a largely forgotten philosophical program that attempts to include Euler's quasi-exploratory activities within a more tolerant conception of "proper mathematics." These views were articulated in the early nineteenth century by a group of British mathematicians to rationalize the innovations in reasoning technique that were swiftly altering algebraic practice in a manner closely analogous to the projective geometry extensions discussed in Section 1. To do so, they divided the study of algebra into two sectors: the "arithmetical" and the "symbolic." As such, the view is rather crudely simplified, but the school frequently provided vivid articulations of the notion that a proper conception of mathematics should

---

[18] Of this passage, Alexander MacFarlane comments: "This is the true phenomenon in algebra; it is more logical than its framer. How can it be possible, unless the algebraist founds his analysis upon real relations? It is the logic of real relations which may outrun the imperfect definitions and principles of the analyst, and make it necessary for him to return to revise them" (1899, pp. 24–25).

[19] I have not found any source for this popular citation earlier than Mach (1895, p. 196).

explicitly embrace its exploratory aspects. George Peacock offers an early articulation of the view:

> Symbolical algebra adopts the rules of arithmetical algebra but removes altogether their restrictions; thus symbolical subtraction differs from the same operation in arithmetical algebra in being possible for all relations of value of the symbols or expressions employed. All the results of arithmetical algebra which are deduced by the application of its rules, and which are general in form though particular in value, are results likewise of symbolical algebra where they are general in value as well as in form; thus the product of am and an which is am+n when m and n are whole numbers and therefore general in form though particular in value, will be their product likewise when m and n are general in value as well as in form; the series for (a+b)n is determined by the principles of arithmetical algebra when n is any whole number; if it be exhibited in a general form, without reference to a final term, may be shown upon the same principle to the equivalent series for (a+b)n when n is general both in form and value. (1830, I, p. vi)

Peacock cites a "principle of the permanence of equivalent forms" as motivational basis, which he explicates as follows: "Whatever algebraical forms are equivalent when the symbols are general in form, but specific in value, will be equivalent likewise when the symbols are general in value as well as in form" (1830, II, p. 59). This obviously represents an attempt to articulate an algebraic analog to Poncelet's "permanence of geometric relationships." Philip Kelland and P. G. Tait amplify upon Peacock's tactics as follows:

> The Algebraic sciences are based upon ordinary arithmetic, starting at first with all its restrictions, but gradually freeing itself from one and another, until the parent science scarcely recognizes its offspring. ... A totally new view of the process of multiplication has insensibly crept in by the advance from whole numbers to fractions. So new, so different is it, that we are satisfied that Euclid in his logical and unbending march could never have attained to it. It is only by standing loose for a time to logical accuracy that extensions in the abstract sciences – extensions at any rate which stretch from one science to another – are affected. ... We trust, then, it begins to be seen that sciences are extended by the removal of barriers, of limitations, of conditions, on which sometimes their very existence seems to depend. Fractional arithmetic was an impossibility so long as multiplication was regarded as abbreviated addition; the moment an extended idea was entertained, ever so illogically, that moment fractional arithmetic started into existence. Algebra, except as mere symbolized arithmetic, was an impossibility so long as the thought of subtraction was chained to the requirement of something adequate to subtract from. ... This is the first idea we want our reader to get a firm hold of; that multiplication is not necessarily addition, but an operation self-contained, self-interpretable – springing originally out of addition; but, when full-grown, existing apart from its parent.

The second idea we want our reader to fix his mind on is this, that when a science has been extended into a new form, certain limitations, which appeared to be of the nature of essential truths in the old science, are found to be utterly untenable; that it is, in fact, by throwing these limitations aside that room is made for the growth of the new science. . . . In the advance of the sciences the old terminology often becomes inappropriate; but if the mind can extract the right idea from the sound or sight of a word, it is part of wisdom to retain it. And so all the old words have been retained in the science of Quaternions to which we are now to advance. (1872, pp. 2–3)

They then discuss how William Rowan Hamilton managed to develop "numbers" (his so-called quaternions) that are capable of capturing three-dimensional machine movements within an algebraic framework (*vide* our earlier discussion of such an evaluative project in Section 3.) The key to his eventual success lay in abandoning an earlier algebraic "necessity" while keeping most its computational surroundings intact:

Hamilton, who early found that his road was obstructed – he knew not by what obstacle – so that many points which seemed within his reach were really inaccessible. He had done a considerable amount of good work, obstructed as he was, when, about the year 1843, he perceived clearly the obstruction to his progress in the shape of an old law which, prior to this time, had appeared like a law of common sense. (1872, p. 5)

That "old law" was the commutativity of multiplication. De Morgan, in his typical amusing fashion, recasts these developmental processes as a fruitful form of amnesia:

[To frame] a clear idea . . . of this separation of symbolic from arithmetical reasoning, we shall propose the following illustration. A person who has thoroughly studied the algebra of positive and negative quantities, is attacked by a severe illness, on recovering from which he finds all memory of connection between his conceptions and the symbols which represented them totally gone, while his expertness in performing the mere transformations with which algebra abounds remains undiminished. When he sees (a + b)2, he perfectly remembers that its substitute was a2 + 2ab + b2, but what a, b, +, &c. stood for, or might have been supposed to stand for, he has wholly forgotten.

He is now a purely symbolical algebraist. Suppose that he endeavors to recover the meaning of his symbols by close examination of their relations. He remembers, for example, that a + b had such a meaning as made it identical with b +a, and he tries all meanings which will satisfy this condition, and attempts to give conformable meanings to other symbols, in the hope of picking out a set of definitions which shall be consistent with each other, and of which the relations which live in his memory shall be logical consequences. He succeeds in his attempt, and thus gives meaning to his transformations and converts his symbolical algebra into a deduction from some fundamental notions of magnitude which he has slowly recovered.

Perhaps the reader will say, he must then have discovered or remembered that a, b, c, etc. stand for numbers, that + and − mean addition and subtraction, etc., etc. By no means; the tenor of this article will require us to show another set of meanings on which he may have happened to alight, not only as consistent with each other as the arithmetical meanings, but more consistent; and [we can further demonstrate] the pliability of the algebraical system, by pointing out that the number of different interpretations under which its symbolical relations will represent truths, is absolutely unlimited. (1840, p. 133)

As I understand these writers (and Dedekind), they evince an appreciation of what we might call *the cultivation and perfection of pure recipe.* We repeatedly find ourselves following analogous reasoning sequences (task 1 → task 2 → ... → task n) within applications that otherwise scarcely resemble one another. Writers of the time were much struck by what they called "mechanical analogies" – the fact that the steps required to reason about an electrical circuit are structurally very similar to how we should reason about a mechanism assembled from springs and dashpots. Following Peacock's "symbolical" suggestions, George Boole discovered that he could successfully resolve difficult logic problems (if they were solvable at all) by utilizing an algorithmic routine markedly similar to how he would approach a high school algebra problem; he merely needed to fiddle with the local rules a little bit along the way (e.g., calculate "1 + 1" as " 0"). In doing so, to be sure, he frequently found his computations passing through uninterpretable stages similar to the "impossible numbers" that Cardano had confronted. Boole drew exactly this same comparison himself:

> The principle in question may be considered as resting upon a general law of the mind, the knowledge of which is not given to us a priori, that is, antecedently to experience, but is derived, like the knowledge of the other laws of the mind, from the clear manifestation of the general principle in the particular instance. A single example of reasoning, in which symbols are employed in obedience to laws founded upon their interpretation, but without any sustained reference to that interpretation, the chain of demonstration conducting us through intermediate steps which are not interpretable to a final result which is interpretable, seems not only to establish the validity of the particular application, but to make known to us the general law manifested therein. No accumulation of instances can properly add weight to such evidence. It may furnish us with clearer conceptions of that common element of truth upon which the application of the principle depends, and so prepare the way for its reception. It may, where the immediate force of the evidence is not felt, serve as a verification, a posteriori, of the practical

validity of the principle in question. But this does not affect the position affirmed, viz., that the general principle must be seen in the particular instance – seen to be general in application as well as true in the special example. The employment of the uninterpretable symbol $\sqrt{-1}$ in the intermediate processes of trigonometry furnishes an illustration of what has been said. (1854, p. 62)[20]

This is rather murky, but the general idea is that, as human reasoners, we find ourselves natively equipped with various "general rules of the mind" (viz., adaptable reasoning routine) that can be harnessed to a wide array of purposes after a suitable degree of specialized tinkering. As mathematicians, we should sometimes attempt to perfect the general recipe as efficiently as we can, without worrying excessively about its potential applications in doing so. Peacock's division between "arithmetic" and "symbolic" studies is founded upon this same bifurcation in investigative labor. And we have already witnessed allied recommendations in Dedekind. In his theory of ideals, he is eager to articulate the general steps required to enlarge an awkward initial algebra σ^* into a sleeker and potentially more informative factoring domain σ^*. However, he leaves to others the task of deciding whether these reasoning improvements will prove of any benefit within a concrete application. He merely ensures that the tactics of evaluator-object supplementation are available to other reasoners if they find that such "logical" enhancements prove helpful within an application.

The recommended "division in investigative labor" will become crucial in our final section, when we return to the developmental tensions between "certainty" and "innovation."

Through these recipe-inspired tactics, our pencils often "prove wiser than ourselves," and we may receive surprising "suggestions from the symbols" that we could have never fancied otherwise. John Venn characterizes these Eulerian opportunities as follows:

> We might conceive the symbols conveying the following hint to us: Look out and satisfy yourselves on logical grounds whether or not there be not an inverse operation to the above. If you can ascertain its existence, then there is one of our number at your service to express it. In fact, having chosen one of us to represent your logical analogue to multiplication, there is another which you are bound in consistency to employ as representative of your logical analog to its inverse, division, – supposing such an operation to exist. (1894, p. 74)

Here Venn codifies an important methodological lesson that contemporary philosophers of method frequently overlook. Sometimes the very circumstances that lead our mathematical models to break down in a conventional manner turn

---

[20] I discuss the close affinities between Boole and Heaviside in (Wilson 2006).

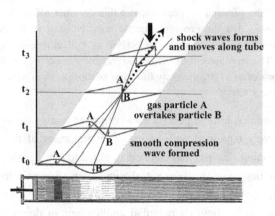

Figure 13. Shock wave formation

out to supply the descriptive apparatus we require to capture the target phenomena in the most efficient manner possible. In a conventional sense, a "blowup singularity" represents a position where some of the quantities within one's modeling equations are forced to take infinite or inconsistent values.

A celebrated example arises with respect to the motion of air within a one-dimensional tube. If we drive a strong enough pulse into the tube via a plunger, the natural modeling equation dictates that the ensuing packet of air will accelerate according to the force with which it has been pushed from behind. After a certain period of time, a "blowup" inconsistency appears in the developing model, analogous to a traffic pileup when the cars from behind travel faster than the cars in front. (This is a common behavior within nonlinear circumstances.) Mathematically, these breakdowns represent situations where our modeling assumptions assign inconsistent pressures to the same packet of air. We might normally expect to abandon our original modeling in disgust when this happens, but Bernhard Riemann recommends otherwise. We can plot the exact location of this "impossibility" perfectly and further calculate how it should progressively move forward along the tube – it constitutes what we normally call a "shock wave." So it turns out the descriptively most valuable aspect of our gas modeling lies in the manner in which its internal devices *fail* (supplying an unusual instance of "an exception that proves the rule"). Such a "cultivation of singularities" comprises an important ingredient within many of Riemann's astonishing innovations.

Boole's conception of methodology directly inspires the equally striking departures from Weierstrassian rigor that appear within the strange "operational calculus" devised by the self-educated engineer Oliver Heaviside. He reached valuable results within electrical engineering by indulging in the most whimsical of calculus manipulations (e.g., "dividing" by the differential operator 1/dt). Earlier we cited his exasperated response to the "Cambridge mathematicians," who only offer the "wet blanket of the rigorists" and refuse "to wander about and be guided by circumstance in the choice of paths." (2003, II, p. 3). He provides an equally vivid defense of his quasi-inductive explorations:

> It is by the gradual fitting together of the parts of a distinctly connected theory that we get to understand it, and by the revelation of its consistency. We may begin anywhere, and go over the ground in any way. ... It may be more interesting and instructive not to go by the shortest logical course from one point to another. It may be better to wander about and be guided by circumstance in the choice of paths, and keep our eyes open to the side prospects, and vary the route later to obtain different views of the same country. Now it is plain enough when the question is that of guiding another over a well-known country, already well explored, that certain distinct routes may be followed with advantage. But it is somewhat different when it is a case of exploring a comparatively unknown region, containing trackless jungles, mountains and precipices. To attempt to follow a logical course from one to another would then perhaps be absurd. ... You have first to find out what there is to find out. ... Shall I refuse my dinner because I do not fully understand the processes of digestion? No, not if I am satisfied with the result. ... First, get on, in any way possible, and let the logic be left for later work. (2003, II, p. 2, and III, p. 370)

As it happens, it took nearly sixty years before other mathematicians puzzled out a sensible underlying "logic" to support Heaviside's strange manipulations.

The Heaviside case is quite extreme in its rejection of conventional rigor, but quasi-empirical factors frequently redirect mathematical inquiry in subtler ways, in the manner in which applied mathematicians patch together disparate ingredients in reaching a sustainable form of physical modeling. Good examples can be encountered among the many variants of boundary value problems that may involve the very same equation (Laplace's) that had occasioned Riemann's difficulties with the Dirichlet Principle. David Hilbert later demonstrated that reasonings similar to Riemann's can be properly implemented if their attendant circumstances support a suitable notion of a "weak limit." The basic challenge is allied to Prym's counterexample: How should a rapidly fluctuating interior match with a piecewise smoother boundary? However, it turns out that there is no universal answer as to how this coupling should be established. When we consult a modern textbook on these topics, we encounter a bewildering array of subtly distinct function spaces, precisely to offer the tools that potentially might be required to complete this alignment task. In particular, subtle facts about energy storage force modeling situations that closely resemble on superficial inspection actually demand that we interpret the "meanings" of our mathematical

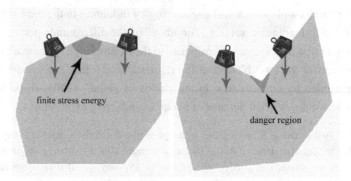

Figure 14. Stresses inside polygons

expressions in a manner that reflects these differences in energetic condition. Employed as a model of a flexible membrane and interpreted naïvely, a 2D Laplace equation equates the film's strain energy storage with its localized degree of bending ($d^2h/dx^2 + d^2h/dy^2$). But there is a wide range of natural situations in which this naïve identification breaks down and a subtler "weak solution" approach is required. Sharp corners along the membrane boundary provide good illustrations of the subtleties required. If the corners in our polygon are all convex (as on the left), the strain energy captured within the finite patches of the membrane surrounding the corner remains finite, and this fact can be exploited to frame a suitable notion of a "weak solution" to our equation following the usual techniques of Schwartz's "theory of distributions."

When we shift to polygons that contain inset corners, as illustrated on the right, the internal fabric becomes twisted in a more drastic manner in which our normal standards of membrane energy storage give way. And these mathematical difficulties correlate with the physical fact that sharp inset corners act as problematic stress concentrators in real life.[21] However, we can use these modeling difficulties for beneficial descriptive purpose if we imitate Riemann's approach to shock waves. For if we can establish that the truly bad spots in our modeling will only arise in localized regions, we have learned valuable data with respect to the problematic stresses. To implement our revised policy of "locating the bad sectors" mathematically, a rather different conception of "weak solution" needs to be framed in which we "cut out" and "jump across" the local patches of material that appear to be asymptotically heading towards a blowup.

The adjusting readings of "energy storage" that we witness in these subtly different recastings recall some of De Morgan's earlier remarks:

---

[21] The Liberty ships constructed during World War II sometimes cracked apart because they were unwisely outfitted with rectangular storage hatches.

[A]nother set of meanings upon which they happened to alight, that are just as harmoniously consistent with each other as were their former meanings, but obtaining a fresh usage within a new arena, due to the fact that the number of different interpretations under which the symbolical relations uncovered through disciplined calculation may turn out to codify represent truths, is absolutely unlimited. (1840, p. 133)

And this is basically the reason why a modern engineering textbook contains so many tools from functional analysis: anticipating which of their subtle combinations is required to suit a potential application is very difficult. To be sure, each component piece of functional analysis equipment ought to be rigorously specified in Hardy's "habit of definitions" sense, but this initial preparation does not override the fact that a good deal of later trial and error assembly may still be required to obtain what Heaviside called "the go" of an effective descriptive routine:

For it is in mathematics just as in the real world; you must observe and experiment to find the go of it. ... All experimentation is deductive work in a sense, only it is done by trial and error, followed by new deductions and changes of direction to fit circumstances. Only afterwards, when the go of it is known, is any formal explication possible. Nothing could be more fatal to progress than to make fixed rules and conventions at the beginning, and then go on by mere deduction. You would be fettered by your own conventions, and be in the same fix as the House of Commons with respect to the dispatch of business, stopped by its own rules. (2003, II, p. 33)

And these same inconstancies of equational "meaning" prompt Hans Lewy to observe:

[Mathematical analysis only supplies] kind of hesitant statements. ... In some ways analysis is more like life than certain other parts of mathematics. ... There are some fields ... where the statements are very clear, the hypotheses are clear, the conclusions are clear and the whole drift of the subject is also quite clear. But not in analysis. To present analysis in this form does violence to the subject, in my opinion. The conclusions are or should be considered temporary, also the conclusions, sort of temporary. As soon as you try to lay down exact conditions, you artificially restrict the subject. (Reid 1991, p. 264)

In the case at hand, we cannot really determine what the Laplace equation "means" until these supplementary issues of energy storage have been resolved, for the correct choice of a "weak reading" for its derivatives hinges upon their resolution.[22]

---

[22] Terence Tao (2010) supplies a "properties" chart like those one finds in design manuals for engineers.

Contemporary philosophical discussion often overlooks Levy's adjustments in equational meanings, which is unfortunate since they are frequently symptomatic of significant shifts in interpretational setting. As a result, many philosophers currently believe that applied mathematicians simply borrow ready-made "structures" from the abundant warehouses of prefabricated pure mathematics (Wilson 2020). Such a point of view overlooks the manner in which a modern engineering textbook opportunistically fashions its equational "meanings" to suit the specific demands of the application at hand.

This same coarse-graining is also reflected in the popular conception of "the mathematics that science needs," as that phrase is employed by W. V. Quine and accepted as coherent within the tremendous range of "naturalist" and "anti-naturalist" literature that Quine's work has spawned. We cannot delve further into these issues, except to remark that they likewise derive from a failure to track the mechanics of effective representation within applied mathematics in adequate nuance and detail.

7 Unsuspected Kinships and Disassociations

> [T]he mathematical facts worthy of being studied are those which, by their analogy with other facts, are capable of leading us to the knowledge of a mathematical law just as experimental facts lead us to the knowledge of a physical law. They are those which reveal to us unsuspected kinship between other facts, long known, but wrongly believed to be strangers to one another.
>
> Henri Poincaré (1908, p. 386)

Viewed retrospectively, the English algebraists' limited conceptions of "numerical" and "symbolic" algebras cannot adequately account for the wide menagerie of innovative "domains-unto-themselves" spawned over the long years of mathematical development. In his early writings, Dedekind sometimes stressed "organic development" from preexistent inquiries in an excessively conservative manner:

> In mathematics too, the definitions necessarily appear at the outset in a restricted form, and their generalization emerges only in the course of further development. But – and in this mathematics is distinguished from other sciences – these extensions of definitions no longer allow scope for arbitrariness; on the contrary, they [must] follow with compelling necessity from the earlier restricted definitions, provided one applies the following principle: Laws which emerge from the initial definitions and which are characteristic for the concepts that they designate are to be considered as of general validity. Then these laws conversely become the source of the generalized definitions if one asks: How must the general definition be conceived in order that the discovered characteristic laws be always satisfied?
> (1854, p. 757)

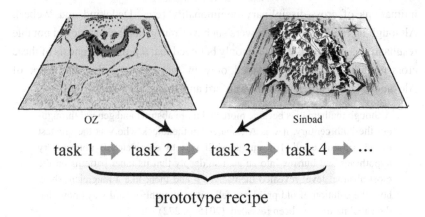

task 1 ➡ task 2 ➡ task 3 ➡ task 4 ➡ ⋯

prototype recipe

Figure 15. Recipe similarity

However, fresh conceptualization also enters mathematics through the "unsuspected kinships" that Poincaré had in mind, in which the innovative elements emerge through the cross-fertilization of distinct modes of inquiry rather than the single-minded development of any specific theme. We can illustrate the distinction by returning to the navigational analogy of Section 1. In many fairy tales, someone is asked to perform a number of difficult tasks, with eventual completions that look entirely different between the various narratives. (Figure 15 compares the seventh voyage of Sinbad with one of the Oz books.) But a closer inspection can reveal that in a large number of these stories the protagonists decompose their assigned tasks along a common organizational chart similar, if rather abstract, procedures (e.g., "arrange your schedule so that nearby tasks are accomplished at more or less the same time").

In his later work, Dedekind became greatly impressed by the fact that similar reasoning policies prove useful in dealing with both number systems and algebraic fields, without there being any evident surface reason why this should be so. Many of his contemporaries felt the same way. Thus J. J. Sylvester proclaims:

> Time was when all the parts of the subject were dissevered, when algebra, geometry, and arithmetic either lived apart or kept up cold relations of acquaintance confined to occasional calls upon one another; but that is now at an end; they are drawn together and are constantly becoming more and more intimately related and connected by a thousand fresh ties, and we may confidently look forward to a time when they shall form but one body with one soul. (1870, p. 262)

Some of the considerations that affected his tactics with respect to the manner in which he developed his theory of ideals reflect an attempt to codify these hazy intimations of cross-disciplinary commonality (e.g., Dedekind and Weber). Although he (and his later followers such as Emmy Noether) achieved notable results in this direction, it is commonly believed that a proper diagnosis of these cross-field affinities was obtained only in the 1950s, within the work of Alexander Grothendiek. David Mumford and John Tate write:

> Although mathematics became more and more abstract and general throughout the 20th century, it was Alexander Grothendieck who was the greatest master of this trend. His unique skill was to eliminate all unnecessary hypotheses and burrow into an area so deeply that its inner patterns on the most abstract level revealed themselves – and then, like a magician, show how the solution of old problems fell out in straightforward ways now that their real nature had been revealed. (2015, p. 272)

Articulating these abstract commonalities frequently requires a conceptual flexibility that is not easily accommodated by the familiar tools of logicist construction. In our earlier discussion of von Staudt, we noted that evaluative properties can provide directive guidance similar to that of a landmark object, and two recipes for geometrical construction can prove essentially the same except that the first draws two lines through a concrete point while the second asks the lines to remain parallel to one another. This is also why I considered the "evaluator-objects" within one of Dedekind's sheltered domains as hybrid in character, simultaneously displaying function- and object-like characteristics. In wider contexts, we sometimes require more liberal forms of cross-grammatical affinity. For example, in formulating general theorems precisely, we frequently must set aside various annoying exceptions. In some applications, this task proves relatively easy – for we can accommodate the exceptions with a simple "except for a set of measure zero" exclusion (which we can normally codify with some quantifiers and a bit of set theory). But in geometry, the allied problems become trickier because of a need for "generic lines." Consider a figure-eight E drawn over the real numbers. Intuitively, E should be classified as being of degree four because an arbitrary line should cut through E in four places. But certain "exceptional lines" fail to do this (either by glazing E along a tangent or passing through E's center). Traditionally, these putative counter-examples were tamed through appeal to "generic lines" – that is, "every generic line intersects E in four places."

Implementing these exclusionary provisos in a precise manner encounters difficulties when we wish to amalgamate previously independent systems algebraically. If we superimpose a further curve C upon our figure-eight E, the generic lines appropriate to C + E may not relate in any simple way to those for C and E individually. However, in other appearances of the same basic combinatorial recipe, the corrective exclusions may look entirely different. To properly capture the cross-disciplinary affinities between recipes, mathematicians sometimes devise odd categories to accommodate the varied exceptions. In this vein, modern algebraic geometers convert their own "almost all" quantifications into a novel form of singular "object" ("*the* generic line") that hovers alongside our figure-eight E. (I have sometimes seen the analogous "generic point" construction pictured as a fuzzy ball hovering alongside E.) I once wrote an article called "Can We Trust Logical Form?" (the implied answer was "no") to highlight these shifting pressures upon grammatical categorization. It supplies a further reason for confining our effective algebras to "spaces" of their own; we cannot tell in advance whether a crucial stage within a common recipe will manifest itself as a concrete gnome or merely as a concept-like gnome-evaluator.

Mathematicians sometimes characterize these classificatory ambiguities as "informal ideas." V. I. Arnold cites J. J. Sylvester on the manner in which renewed examinations of subterranean strategy can affect the parsing of established mathematical claims:

> Sylvester already described as an astonishing intellectual phenomenon, the fact that "general statements are simpler than their particular cases." The anti-Bourbakist conclusion that he drew from this observation is even more striking. According to Sylvester, "a mathematical idea should not be petrified in a formalised axiomatic setting, but should be considered instead as flowing as a river." One should always be ready to change the axioms, preserving the informal idea. (2000, p. 404)

Wittgenstein appears to offer a similar comment:

> It might be imagined that some propositions, of the form of empirical propositions, were hardened and functioned as channels for such empirical propositions as were not hardened but fluid; and that this relation altered with time, in that fluid propositions hardened, and hard ones became fluid. The mythology may change back into a state of flux, the river-bed of thoughts may shift. But I distinguish between the movement of the waters on the river-bed and the shift of the bed itself; though there is not a sharp division of the one from the other. (1969, §§96–97)

As I understand his drift, at a particular developmental stage a suitable codification of the stages we want to pass through in applying a well-established reasoning recipe will seem as if they are optimally registered within a rule-driven general "calculus" of type χ. These strictures will normally serve as the guiding beacons of our everyday calculations. Nonetheless, the shifting winds of innovative evolution we have studied may eventually recommend some alternative set of leading principles χ^*.

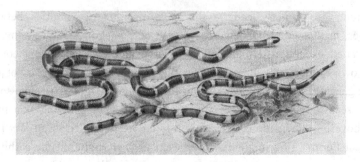

Figure 16. Biological mimics

Indeed, careful reexamination sometimes reveals that reasoning tasks that appear similar on the surface actually differ in substantial respects. (Elsewhere, I have called these "semantical mimics," analogous to the manner in which the innocuous corn snake at the bottom of Figure 16 closely resembles its ferocious coral snake companions.) My favorite exemplar of this "semantic mimicry" can be found in the manner Newton's celebrated decomposition of light into colors superficially resembles the straightforward Fourier analysis applicable to sound waves. In such cases, the interior content of Newton's reasonings must undergo a process that geologists call "replacement remineralization": the shape of fossilized traces remains constant, but their chemical contents become completely altered.

The sometimes acrimonious squabbles between set theorists and category theorists over "the foundations of mathematics" can be amiably defused in this light. The latter school seeks the descriptive vocabularies that can best capture the affinities between distinct forms of reasoning recipe, whereas the set theorists are more centrally concerned to ensure that the truth-values of within a novel domain have been adequately specified (the specific purpose to which Dedekind applies his ideals). The tools that facilitate these latter purposes (e.g., equivalent classes of the form $\{x|Rxa\}$) prove unhelpfully clunky if they are expected to capture the cross-disciplinary affinities for which the discriminations of category theory are better suited. Fights break out when either party exclusively claims the mantle of "foundations of mathematics" only to itself.

Insofar as the objectives of this Element goes, we only need to acknowledge that issues of "hidden kinship" can also "organically drive" mathematical development in innovative directions inadequately acknowledged by the original projective geometers and English algebraists who concentrated too exclusively upon the advantages accruing from extension element supplements.

8 The Enlarging Architecture of Mathematical Reasoning

The buildings of science are not erected the way a residential property is, where the retaining walls are put in place before one moves on to the construction and expansion of living quarters. Science prefers to get inhabitable spaces ready as quickly as possible to conduct its business. Only afterwards, when it turns out that the loosely and unevenly laid foundations cannot carry the weight of some additions to the living quarters, does science get around to support and secure those foundations. This is not a deficiency but rather the correct and healthy development.

David Hilbert (Corry 1999, pp. 163–64)

In this concluding section, let us attempt to consolidate our Dedekind-influenced musings into a coherent reply to our central "innovation versus certainty" dilemma. To these purposes, we begin with a lengthy but revealing passage from his early habilitation lecture:

If one finds the chief goal of each science to be the endeavor to fathom the truth – that is, the truth which is either wholly external to us, or which, if it is related to us, is not our arbitrary creation, but a necessity independent of our activity – then one declares the final results, the final goal (to which one can in any case usually only approach) to be invariable, to be unchangeable. In contrast, science itself, which represents the course of human knowledge up to these results, is infinitely manifold, is capable of infinitely different representations. This is because, as the work of man, science is subject to his arbitrariness and to all the imperfections of his mental powers. There would essentially be no more science for a man gifted with an unbounded understanding – a man for whom the final conclusions, which we attain through a long chain of inferences, would be immediately evident truths; and this would be so even if he stood in exactly the same relation to the objects of science as we do. This diversity in the conceptions of the object of a science finds its expression in the different forms, the different systems, in which one seeks to embed those conceptions. We can see this everywhere. For instance, in the descriptive natural sciences we see it in the grouping and classification of materials. Depending on the greater or lesser importance which the investigator of nature attaches to a criterion as a concept suited to distinguishing and classification, he either elevates it to a chief touchstone of classification, or else he uses it to designate merely inessential differences. Thus in mineralogy we have two systems, one of which rests on the chemical composition of mineral bodies, and the other on their crystallographic, morphological nature. These two systems clash, and nobody has managed until now to bring them into complete harmony with each other. Each of these systems is perfectly justified, for science itself shows that similar bodies group themselves together most naturally in these ways. But it will occur to no mineralogist to advance, say, differences of color as the characteristic features and to prefer a classification resting on them to all others. Of course, no reason can be given against this a priori, but experience teaches in the course of research that color is not of such great significance for the true nature of bodies as are the criteria mentioned earlier; or, to put it another way, experience teaches that one cannot operate so certainly, so effectively, with color as with the other criteria. The introduction of such a concept as a motif for the arrangement of the system is, as it were, a hypothesis which one puts to the inner nature of the science; only in further development does the science answer; the greater or lesser effectiveness of such a concept determines its worth or worthlessness. (1854, pp. 755–56)

As I understand Dedekind's intent here, it centers upon the observation that *productive reasoning recipes* can be cultivated in their own right as a special component within methodological inquiry, set somewhat apart from whatever direct utilities they may display within a specific form of application. Indeed, a discerning probing of "How do these strategic recipes operate?" will typically unfold along markedly different investigative pathways than when we directly attend to their brute applicational successes. (We will consider some sample illustrations later.) It is also important that we ascertain whether a scheme's formal operations can be certified as standing in good working order. (No trouble spots comparable to a toleration of "6/0" or breakdowns of Dirichlet's principle have been overlooked.) We should further inquire whether our scheme's factoring capacities might be improved through the supplementation of suitable "ideals." And so Dedekind recommends a reasonable division within our investigative labors, for much can be learned of the computational reach of a reasoning recipe without worrying whether the routine will ever serve any practical purpose. In the same manner, a key aspect of Heaviside's "finding out what there is to find out" requires that we frame an accurate appraisal of the formal reach of our reasoning tools before we apply them to physical circumstances. As instructors, we should warn our knights-in-training: "Callow youths, you are not ready to confront fierce opponents. First perfect your skills at home and extend the capacities of your weaponry." When these gallants enter the jousting field later on, they may find themselves deploying their lances and shields in improvisatory manners of which they never dreamed. Nonetheless, their preparatory "cultivation and perfection of computational recipe" will have served them well. As we already noted, the utilities of concentrating upon exploratory skills independently of eventual purpose lie at the center of the English algebraists' emphasis upon training within their "symbolical algebra."

Similar views are encountered among many of Dedekind's contemporaries. For example, the remarkable author/entrepreneur Paul Carus explicitly allocates the investigation of reasoning per se to an intermediate epistemological status between the wholly a priori and the wholly empirical:

Superstitions are much more common than is generally assumed, for they not only haunt the minds of the uneducated and uncivilized, but also those of the learned. Science is full of superstitions, and one of the most widespread of its superstitions is the belief in axioms.

"Axiom" is defined as "a self-evident truth."

It is not the peasantry who believe in axioms, but some of the most learned of the learned, the mathematicians; and since mathematics, with all its branches, is a model science, the solid structure of which has always been admired and envied by the representatives of other sciences, so that they regarded it as their highest ambition to obtain for the results of their own investigations a certainty equal to the certainty of mathematical arguments. . . . At present it is generally supposed that we have to accept either the one or the other horn of this dilemma: either axioms are the result of an elaboration of particular experiences, i. e., are, like all other know-ledge concerning the nature of things, a posteriori, or they are conditioned by the nature of human reason, they are a priori. The most prominent representative of the former view is John Stuart Mill; of the latter, Kant. (1893, p. 3752)

He attempts to ground these intermediate studies within the *de facto* psychology of human conceptual creativity, rather than appealing to some supportive external "ontology":

Our solution of this perplexing problem is to regard the rules of reasoning, such as Euclid has formulated under the name of postulates, as products of rigidly formal reasoning. Man's reasoning consists of his mental operations, and man's mental operations are acts. The mere forms of mental acts are such as advancing step by step from a fixed starting-point. We thus create purely formal magnitudes . . . Purely mental acts are, as acts, not different from any other happenings in the world. The sole difference consists in their being conscious, and that for convenience sake a starting-point is fixed as an indispensable point of reference. . . .

Our mental operations, by which the rigidly formal products, commonly called a priori, are produced, being the given data out of which mind grows, and as regards their formal nature being the same as any other operations in the world, we say that the products of these operations are ultimately based upon experience. However, they are not experience in the usual (i.e., Kant's) sense of the word; they are not information received through the senses. They are due to the self-observation of the subject that experiences, and this self-observation is something different from the mysterious intuition in which the intuitionists believe. The subject that experiences does not take note of external facts, but of its own acts, constructing general schedules of oper-ations which hold good wherever the same operations are performed.

Thus on the one hand we deny that the rigidly formal truths are general-izations abstracted from innumerable observations; and on the other hand that they are axioms or self-evident truths, or principles acquired by some kind of immediate intuition. We recognize their universality and necessity for all kinds of operations that take place, and yet escape the mysticism that our

surest and most reliable knowledge must be taken for granted, that it is
unproved, unprovable and without any scientific warrant. (1893, p. 3754)

Like Dedekind, Carus believes that these mathematical investigations of
productive reasoning should progressively enlarge by emerging "organically"
from epistemological grounds already gainsaid, rather than pursuing the
unmotivated and willy-nilly permissions permitted within "anything goes" if-
thenism. By emphasizing the "organically evolved" aspects of our most significant specimens of mathematical innovation, Carus avoids the unhappy shoals of an
undesirable Platonic mysticism by instead associating mathematics' "nonempiri-
cal" character with the pragmatic utilities of its wide-ranging investigations. These
can enhance the reliability of our reasoning tools across a wide range of topics,
many of which may have no bearing on physical applications at all. This distan-
cing-from-the-empirical reflects a prudent division of labor with respect to investigative mission – we should verify that the internal mechanics of a proposed system
of algebraic rules are well-constructed without gaps before we bother presenting
them "to the inner nature of a science" as an organizational "motif." According to
Carus's diagnosis, this methodological sheltering from the directly empirical
supplies mathematical inquiry with the surface patina of unchallenged "certainty"
that Platonists mistake for a direct grounding within an elusive ontological realm.

Hermann Lotze offers roughly parallel sentiments, likewise based upon our
inherent limitations as human reasoners:

> Only a mind which stood at the center of the real world, not outside individual
> things but penetrating them with its presence, could command such a view of
> reality as left nothing to look for, and was therefore the perfect image of it in
> its own being and activity. But the human mind, with which alone we are here
> concerned, does not thus stand at the center of things, but has a modest
> position somewhere in the extreme ramifications of reality. Compelled, as it
> is, to collect its knowledge piecemeal by experiences which relate immedi-
> ately to only a small fragment of the whole, and thence to advance cautiously
> to the apprehension of what lies beyond its horizon, it has probably to make
> a number of circuits, which are immaterial to the truth which it is seeking, but
> to itself in the search are indispensable. However much, then, we may
> presuppose an original reference of the forms of thought to that nature of
> things which is the goal of knowledge, we must be prepared to find in them
> many elements which do not directly reproduce the actual reality to the
> knowledge of which they are to lead us: indeed there is always the possibility
> that a very large part of our efforts of thought may only be like a scaffolding,
> which does not belong to the permanent form of the building which it helped
> to raise, but on the contrary must be taken down again to allow the full view of
> its result. (1888, p. 9)

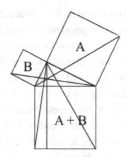

Figure 17. Euclidean proof

In referring to the "indispensable circuits, which are immaterial to the truth which it is seeking," Lotze appears to have in mind the mental "scaffolding" to which the astronomer John Herschel appealed in an earlier essay:

> In erecting the pinnacles of this temple [of science], the intellect of man seems quite as incapable of proceeding without a scaffolding or circumstructure foreign to their design, and destined only for temporary duration, as in the rearing of his material edifices. A philosophical theory does not shoot up like the tall and spiry pine in graceful and unencumbered natural growth, but, like a column built by men, ascends amid extraneous apparatus and shapeless masses of materials. (1857, p. 67)

Euclid's celebrated proof of the Pythagorean theorem nicely illustrates what Herschel had in mind. (Figure 17) Directly inspecting the situation of a right triangle with squares erected on its sides does not convey a reliable conviction that these areas add up in the required manner, but once we encase the figure inside Euclid's quiver of superadded lines, the conclusion becomes forced upon us as unavoidable. Is it not remarkable that we somehow manage to gain greater certitude through a seemingly extraneous embedding of this sort? (Kant certainly thought so.) In fact, Arthur Schopenhauer found the procedure so odd that he complained of "the brilliant perversity" of Euclid's proof techniques:

> If we now turn ... to mathematics, as it was established as a science by Euclid, and has remained as a whole to our own day, we cannot help regarding the method it adopts, as strange and indeed perverted. ... Instead of giving a thorough insight into the nature of the triangle, [Euclid] sets up certain disconnected arbitrarily chosen propositions concerning the triangle, and gives a logical ground of knowledge of them, through a laborious logical demonstration, based upon the principle of contradiction. Instead of an exhaustive knowledge of these space-relations we therefore receive merely certain results of them, imparted to us at pleasure, and in fact we are very much in the position of a man to whom the different effects of an ingenious

machine are shown, but from whom its inner connection and construction are withheld. We are compelled by the principle of contradiction to admit that what Euclid demonstrates is true, but we do not comprehend why it is so. We have therefore almost the same uncomfortable feeling that we experience after a conjuring trick, and, in fact, most of Euclid's demonstrations are remarkably like such feats. The truth almost always enters by the back door, for it manifests itself per accidens through some contingent circumstance. Often a reducio ad absurdum shuts all the doors one after another, until only one is left through which we are therefore compelled to enter. Often, as in the proposition of Pythagoras, lines are drawn, we do not know why, and it afterwards appears that they were traps which close unexpectedly and take prisoner the assent of the astonished learner, who must now admit what remains wholly inconceivable in its inner connection, so much so, that he may study the whole of Euclid through and through without gaining a real insight into the laws of space-relations, but instead of them has only learns by heart certain results which follow from them. This specially empirical and unscientific knowledge is like that of the doctor who knows both the disease and the cure for it, but does not know the connection between them. (1909, pp. 90–92)

Schopenhauer hoped that a more direct consultation of "inner intuition" in a quasi-Kantian manner might provide a more direct "insight into space-relations." But this supposition was, in fact, erroneous. The "alternative proofs" he favored typically exploit movements of components in the manner illustrated in Figure 18 (which, indeed, strikes many of us as more appealing psychologically). However, we have also learned from bitter experience to distrust verification tactics of this stripe, for wider experience has such demonstrated that "proofs" of this second character are sometimes deceptive.[23]

Why must we check our reasoning policies in these roundabout ways? An appropriate answer must begin with the basic limitations and biases inherent within the core capacities for reasoning that we have inherited from our hunter-gatherer ancestors. We find that we often cannot reliably evaluate the

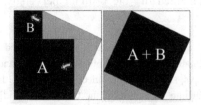

Figure 18. Movement proof

---

[23] A beautiful exemplar is Paul Curry's dissection paradox, provided in Gardner (1956, pp. 139–50). Allegedly, the Greeks banned appeals to motion in their demonstrations for essentially these reasons.

trustworthiness of the conclusions we reach with respect to topic X until we have examined how such inquiries behave when transferred to a parallel setting Y of a disparate character. Given two curves C and C', how will they intersect when placed on top of one another? A successful answer requires that the interior characteristics of that little dimple at p on C be scrutinized closely. The best way to do this is to blow up the point within a space of higher dimensions in which the hidden complexities sheltered within p become disentangled in a more humanly assessable manner (this tactic is called a "resolution of singularities"). Our mathematical heritage has developed a large array of sophisticated tests for gauging the potential reliabilities and defects of our reasoning tools, often in ways that carry us into totally unexpected domains.

Until our inferential skills have been subjected to critical inquiry of this variegated nature, we generally will not be able to reliably determine what we have captured within our physical vocabularies either. Consider the familiar notion of a "light frequency" in optics. By decomposing natural light into these components, our practical capacities to deduce valuable predictive results become enormously enhanced. Yet such successes alone do not ensure that we actually know what we have been talking about referentially. Indeed, the physicist's understanding of what "frequency" signifies has been subjected to many astonishing reversals in interpretation over its developmental history. Students of optics are usually startled when they eventually encounter blithe "corrections" in their advanced textbooks, such as: "The monochromatic sources and monochromatic fields discussed in most optics textbooks are not encountered in real life" (Wolf 2007, p. xi). Such revisionary appraisals are often driven by matters of mathematical coherence with respect to reasoning recipe, such as the considerations that led Lord Rayleigh to proclaim:

> [I]t may be well to emphasize that a simple vibration implies infinite continuance, and does not admit of variations of phase or amplitude. To suppose, as is sometimes done in optical speculations, that a train of simple waves may begin at a given epoch, continue for a certain time it may be a large number of periods, and ultimately cease, is a contradiction in terms. (1894, §65)

As a result, significant conceptual adjustments are required to render standard types of optical specification descriptively coherent. Uncovering the true physical references hiding behind the "scaffolding" featured in the diagrams of our elementary textbooks represents a significant diagnostic task, often involving a good deal of unanticipated mathematical sophistication.

The long and short of all of this is that framing trustworthy improvements upon our inherited collection of reasoning policies represents a delicate and complex intellectual task, in which we find ourselves driven to consider what arises within recondite corners of mathematics far removed from our original foci. But this discrepancy arises with respect to virtually any form of reasoning recipe. A Schopenhauer indicates, a vast gulf separates knowing how to execute a card trick and properly understanding why it works. Even in the case of parlor games, a proper appreciation of when a routine will lead to satisfactory results, and when it will not, usually demands that we examine its procedures within sophisticated mathematical settings that we could not have anticipated beforehand.[24]

The chief reason why we must cautiously utilize these mixed and indirect modes of empirical and mathematical examination stem from Lotze's observation that "we do not stand at the center of things." We have not been decked out a priori with internal concepts whose manner of attachment to the external world can be immediately certified. Instead, we must grope our way to tighter attachments by tinkering with the wide varieties of potential reasoning routine until we happen upon the winners. Sometimes these investigations will lead to clashes of the sort that Dedekind has in mind when he writes of his two schemes for mineralogical classification. Each may seem equally worthy on their own merits, but their conflicting cross-classifications demand a deeper probe of their underlying workings. Obtaining an accommodating reconciliation in this manner is not guaranteed, but it is only through such efforts that we obtain an improved grip upon an external world that is otherwise hard to fathom.

To be sure, the authors cited all hint at additional themes with respect to what Jacques Hadamard called "the psychology of invention in the mathematical field," but I think that we can safely set these supplementary remarks aside. When Dedekind writes that "we are a divine race," he probably alludes to the then-popular thesis that the creative human intellect displays a special ability to "extract the latent generality from the particular," a creative form of mental capacity that sets homo sapiens apart from other animals.[25] But these misty intimations of unconscious insight need not

---

[24] For an expansion upon these considerations, see "A Second Pilgrim's Progress" in (Wilson 2017).

[25] This phrase derives from Boole (2018). Hermann Lotze expresses similar psychologistic sentiments as well. Dedekind refers favorably to contemporary work by Ernst Friedrich Apelt (probably Apelt 1854). Apelt was a follower of Jakob Fries, and it is worth noting that Gauss himself was an admirer of the latter's writings on methodology, to whom he once wrote:

> I have always had a great predilection for philosophical speculation, and now I am all the more happy to have a reliable teacher in you in the study of the destinies of science, from the most ancient up to the latest times, as I have not always found the desired satisfaction in my own reading of the writings of some of the philosophers. (Herrmann 1994)

The often intimate interconnections between mathematicians and philosophers within this time frame merit closer investigation.

be regarded as vital to the central observation that strategic recipes ought to be developed and cultivated in their own right, apart from any question of immediate empirical application. Likewise, the factors that prompt the initial insights of genius (if they exist at all) need not stem from subconscious perceptiveness, for sometimes a novel recipe is simply the happy product of random experimentation. To be sure, confronted with the repeated inventiveness that James Watt displayed across his gamut of admirably designed contraptions (Section 2), someone like Franz Reuleaux can safely presume that some generalizable procedure must lie latent within these artful creations. But Reuleaux uncovers this as yet unarticulated "Science" by closely examining the details of the inventions Watt produced, not by engaging in murky speculations with respect to Watt's secret cognitive processes. In the same way, subconscious reasoning procedures of some unfamiliar stripe must support the astonishing abilities that certain "mental calculators" evince in performing lengthy computations swiftly and accurately, without conscious awareness of how they have obtained their answers. (Smith 1973) Subsequent researchers have enjoyed considerable success in supplying algorithms that capture search tactics of an allied flavor in the form of articulated rules, despite the fact their cognitive implementation within the prodigies themselves will most likely involve irregular approximations and unexpected shortcuts.

In any event, the likelihood that Watt and such prodigies share much in common from a mental perspective is small. The factors that prompt innovative refinements within our reasoning capacities need not trace to any single creative source. But this is not to say that, within a specific mathematical arena, we cannot supply credible reconstructions of the unconscious factors that led the projective geometers to regard their sundry innovations as "correct" and "expressive of hidden inner pattern." Or the intuitive satisfactions that convinced Watt that he had tinkered his inventions into perfection. Methodologists of concrete mathematics have recently supplied valuable insights into the motivational factors that secretly guided earlier generations to the innovations they adopted. ((Manders 1989) and (Baldwin 2018) provide admirable exemplars.)

If these observations are suitably drawn, then the organically developing alignments between localized patches characteristic of real-life mathematical innovation should be approached in the manner that Wittgenstein characterizes as "anthropological" (*vide* Section 2), in which we first map out the current arrangement of central patches within mathematics' widely stretched fabric and then provide convincing rationales for why, for example, the computational experiences within the algebra φ will naturally prompt a correspondent interest in the neighboring algebra φ^*. The collective results may comprise a "motley" in Wittgenstein's sense,[26] but its evolving innovations can still be credited with a developmental "unity" in the sense that the entire patchwork improves itself by coordinating tactics employed elsewhere in the fabric. Such are the developmental considerations behind Dedekind's and De Morgan's insistence that the corpus of "genuine mathematics" can only be properly appreciated if the factors that drive its historical innovations are duly recognized. The whole may still comprise a motivational "motley," without thereby decomposing into a disconnected hodgepodge of "if-then" pursuits.

The views with respect to innovation outlined in this book stand in stark contrast to the semantic stasis popular within analytic philosophical circles today. There it is confidently presumed that some inner core of "absolute necessity" can be firmly carved out by considering "the propositions that remain true in every possible world." Such views invariably rest upon presumptions of easy referential attachment that our nineteenth-century authors would have dismissed as naïve.[27] They instead presumed that we must gradually creep up on reliable word-to-world attachments through lengthy developmental sequences of partial successes and partial failures until our linguistic activities successfully enmesh some broad sector of external circumstance within a net of certifiable pragmatic entanglement. Christopher Pincock (Pincock 2010, p. 120) has labeled such developmental expectations as "patient scientific realism," in contrast to the immediate gratifications promised within the contemporary semantic presumptions that "an inductively supported scientific theory will straightforwardly tell us what the world beyond is like." But the rugged path to reliable referential attachment is quite demanding, and the accumulating improvements that point to Dedekind's "truths which are wholly external to us" require widely variegated investigations in which the supportive underpinnings of our innovations receive constant scrutiny and comparative reappraisal.

"To everything there is a season" advises Ecclesiastes, and sometimes our developmental circumstances demand that we ascertain whether an algebra φ can be naturally coupled to a well-defined neighboring algebra φ^* in which

---

[26] Consider, "Mathematics is a motley of techniques and proofs" (Wittgenstein 1956, III §46). It may be possible to construct an interpretation of Wittgenstein along the lines suggested here, but I will not attempt to do so. I am instead content to cite Emmy Noether once again: "Es steht alles schon bei Dedekind."

[27] W.V. Quine labels such conventional semantic doctrines as "the myth of the mental museum." Its polar opposite is "meaning deflationism," which likewise promises a strategy for "easy attachment" (viz., none is required). I do not believe that a worthy "patient realist" should accept either doctrine.

unique factorization is possible. As we pursue the potential reasoning advantages that lie along this developmental axis, we need not be especially concerned whether these discoveries will eventually facilitate any external purpose; the business at hand is to ensure that the mechanisms of algebraic manipulation can be rendered fully defined without problematic gaps. While pursuing this recipe-focused division in developmental labor, we can set aside the considerations of "reference" and "external truth" that may trouble us at other times. Returning to the earlier matter of Rayleigh's worries about "light frequency," we can distinguish the question of whether a reasoning algebra exists in which point-originated and plane waves can be coherently linked together via a Fourier transform. The answer is "yes," for Laurent Schwartz eventually fulfilled this tricky task in his theory of "distributions." But his notable successes along this algebraic axis do not fully resolve all of the applicational issues that troubled Rayleigh. Those required investigations of a markedly different character.[28] The recognition of these seasonal divisions in investigative labor strikes me as a common chord that underlies the cautionary observations we extracted from Carus, De Morgan, and Dedekind.

With respect to the basic tensions between "necessity" and "innovation" that we have examined over the course of this Element, we may tentatively conclude that monitored innovation in reasoning recipe maintains the upper hand with respect to the continual reshaping of profitable mathematics, whereas its residual patina of persisting "necessity" largely reflects the advisability of cultivating these component patches in Dedekind's standalone manner for a certain span of time. From this vantage point, we need not regard mathematics' alleged "certainty" as a firm philosophical datum but can instead focus upon the tremendous utilities that its evolved tools offer for improving the qualities of our humanly feasible reasoning schemes. Mathematics' central virtues thereby reside in the manner in which such thinking enhances our abilities to reason productively about a wide range of topics, rather than reflecting some Platonic background unto itself. Accordingly, I do not see any great utility in crediting mathematical thinking with a supportive "ontology" of its own – a term that is better reserved for Dedekind's "truths which are wholly external to us." But in their pursuit, we can cite Justice Holmes once again: "Certainty generally is illusion, and repose is not the destiny of man."

---

[28] I will amplify upon this example in a forthcoming essay "How 'Wavelength' Found Its Truth-Values."

References

Apelt, Ernst Friedrich (1854), *Die Theorie der Induction*, Wilhelm Engelmann, Leipzig.

Arnold, V. I. (2000), "Polymathematics: Is Mathematics a Single Science or a Set of Arts?" in Arnold, V.I., Atiyah, M., Lax, P. and Masur, B., eds., *Mathematics: Frontiers and Perspectives*, American Mathematical Society, Providence.

Baker, H. F. (1923), *Principles of Geometry*, vol. 1, Cambridge University Press, Cambridge.

Baldwin, John T. (2018), *Model Theory and the Philosophy of Mathematical Practice*, Cambridge University Press, Cambridge.

Bell, E. T. (1937), *Men of Mathematics*, Simon and Schuster, New York.

Boole, George (1854), *An Investigation of the Laws of Thought*, Walton and Maberly, Cambridge.

(2018), *The Continued Exercise of Reason*, MIT Press, Cambridge.

Bottazzini, Umberto (1986), *The Higher Calculus*, Springer-Verlag, Cham.

Bottazzini, Umberto, and Gray, Jeremy (2013), *Hidden Harmony – Geometric Fantasies*, Springer, Cham.

Burgess, John (2015), *Rigor and Structure*, Oxford University Press, Oxford.

Cayley, Arthur (1889), "Presidential Address to the British Association, September, 1883," in *The Collected Mathematical Papers of Arthur Cayley*, Cambridge University Press, Cambridge.

Cajori, Florian (1919), *A History of Mathematics*, MacMillan, New York.

Carus, Paul (1893), "Axioms," *Open Court Magazine* 7 (31).

Cellucci, Carlo (2017), *Rethinking Knowledge: The Heuristic View*, Springer, Cham.

Clifford, William Kingdon (1891), *The Common Sense of the Exact Sciences*, Appleton, New York.

Coolidge, J. L. (1934), "The Rise and Fall of Projective Geometry," *American Mathematical Monthly* 41 (4).

Corry, Leo (1999), "David Hilbert: Geometry and Physics: 1900–1915," in J. J. Gray, ed., *The Symbolic Universe*, Oxford University Press, Oxford.

Davis, Phillip (1987), "Impossibilities in Mathematics," in Philip J. Davis and David Park, ed., *No Way: The Nature of the Impossible*, W. H. Freeman, New York.

Dedekind, Richard (1854), "On the Introduction of New Functions in Mathematics," in (Ewald 1996).

(1872), "Continuity and Irrational Numbers," in (Dedekind 1963).

(1877), *Theory of Algebraic Numbers*, John Stillwell, trans., Cambridge University Press, Cambridge.

(1888), "The Nature and Meaning of Numbers," in (Dedekind 1963).

(1888a), "Letter to Weber," in (Ewald 1996).

(1963), *Essays on the Theory of Numbers*, Dover, New York.

Dedekind, Richard, and Weber, Heinrich (2012), *Theory of Algebraic Functions of One Variable*, American Mathematical Society, Providence.

De Morgan, Augustus (1840), "Negative and Impossible Quantities," *Penny Cyclopaedia*, vol. 16, Charles Knight, London.

(1849), *Trigonometry and Double Algebra*, Taylor, Walton, and Maberly, London.

Dieudonné, Jean (2009), *A History of Algebraic and Differential Topology*, Birkhäuser, Basel.

Euler, Leonhard (1761), "Observationes de comparatione arcuum curvarum irrectificabilium," cited in (Merz 1912).

Ewald, William, ed. (1996), *From Kant to Hilbert*, vol. 2, Oxford University Press, Oxford.

Forder, H. G. (1927), *The Foundations of Euclidean Geometry*, Cambridge University Press, Cambridge.

Fortney, J. P. (2018), *A Visual Introduction to Differential Forms and Calculus on Manifolds*, Birkhäuser, Cham.

Frege, Gottlob (1874), "Methods of Calculation based on an Extension of the Concept of Quality," in *Collected Papers on Mathematics, Logic and Philosophy*, Basil Blackwood, Oxford (1984).

Gardner, Martin (1956), *Mathematics Magic and Mystery*, Dover, New York.

Glanvill, Joseph (1661), *The Vanity of Dogmatizing*, Henry Eversden, London.

Goethe, J. W. (1906), *Maxims and Reflections*, T. B. Saunders, trans., MacMillan, New York.

Grassmann, Hermann (1844), *Die lineale Ausdehnungslehre*, Otto Wigand, Leipzig. Translation from (Carus 1893).

Gray, Jeremy (2007), *Worlds Out of Nothing*, Springer-Verlag, London.

Hancock, Harris (1928), "The Analogy of the Theory of Kummer's Ideal Numbers with Chemistry," *American Mathematical Monthly* 35(6).

Hankel, Hermann (1875), *Elemente der Projectivischen Geometrie*, cited in (Merz 1912).

Hardy, G. H. (1956), *Divergent Series*, Oxford University Press, Oxford.

(1967), *A Mathematician's Apology*, Cambridge University Press, Cambridge.

Heaviside, Oliver (2003), *Electromagnetic Theory*, 3 vols., Chelsea, Providence.

Herrmann, Kay (1994), "Jakob Friedrich Fries (1773–1843)," Jenenser Zeitschrift für Kritisches Denken.

Herschel, John Frederick William (1857), *Essays from the Edinburgh and Quarterly Reviews with Addresses and Other Pieces*, Longman, Rees, Orme, Brown, and Green, London.

Hilbert, David (1902), "Mathematical Problems," *Bulletin of the American Mathematical Society* 8(10).

Hill, Thomas (1857), "The Imagination in Mathematics," *North American Review* 85.

Holmes, Oliver Wendell (1897), "The Path of the Law," *Harvard Law Review* 10(457).

Jacobi, Karl (1830), "Letter to Legendre," cited in (Bottazzini 1986).

Katz, Victor J., and Parshall, K. H. (2014), *Taming the Unknown*, Princeton University Press, Princeton.

Kelland, Philip, and Tait, P. G. (1873), *Introduction to Quaternions*, MacMillan, London.

Kendig, Keith M. (1983), "Algebra, Geometry, and Algebraic Geometry," *American Mathematical Monthly* 90 (3).

(2010), *A Guide to Plane Algebraic Curves*, Mathematical Association of America, Providence.

Klein, Felix (1908), *Elementary Mathematics from an Advanced Standpoint: Geometry*, E. R. Hedrick and C. A. Nobel, trans., MacMillan, New York.

(1926), *The Development of Mathematics in the 19th Century*, M. Ackerman, trans., Mathematical Science Press, Brookline.

Knopp, Konrad (1990), *Theory and Application of Infinite Series*, Dover, New York.

Kummer, Ernst (1847), "Über die Zerlegung der aus Wurzeln der Einheit gebildeten complexen Zahlen in ihre Primfactoren," *Crelle's Journal*, 35. Translation from (Hancock 1928).

Locke, John (1979), *An Essay Concerning Human Understanding*, Oxford University Press, Oxford.

Lotze, Hermann (1888), *Logic*, vol. 1, Bernard Bosanquet, trans., Oxford University Press, Oxford.

MacFarlane, Alexander (1899), *The Fundamental Principles of Algebra*, The Chemical Publishing Company, Easton.

(1916), *Lectures on Ten British Mathematicians of the Nineteenth Century*, John Wiley, New York.

Mach, Ernst (1895), *Popular Scientific Lectures*, Open Court, Chicago.

Manders, Kenneth (1989), "Domain Extension and the Philosophy of Mathematics," *Journal of Philosophy* 86(10).

Merz, John Theodore (1912), *A History of European Thought in the Nineteenth Century*, vol. 2, William Blackwood, London.

Monna, A. F. (1975), *Dirichlet's Principle*, Oostheok, Scheltema and Holkema, Utrecht.

Moritz, R. E. (1914), *Memorabilia Mathematica*, MacMillan, New York.

Mumford, David, and Tate, John (2015), "Alexander Grothendieck (1928–2014)," *Nature* 517.

Peacock, George (1830), *A Treatise on Algebra*, 2 vols., J. J. Deighton, Cambridge.

Pincock, Christopher (2010), "Critical Notice: Mark Wilson, Wandering Significance," *Philosophia Mathematica* 18(1).

Poincaré, Henri (1902), *Science and Hypothesis*, G. B. Halstead, trans., Dover, New York.

(1908), *Science and Method*, G. B. Halstead, trans., Dover, New York.

Poncelet, J.-V. (1822), Traité des propriétés projective des figures (1822), cited in (Gray 2007).

Rayleigh, Lord (1894), *Theory of Sound*, MacMillan, London.

Reid, Constance (1991), "Hans Lewy, 1904–1988," in P. Hilton, F. Hirzebruch, and R. Remmert, ed., *Miscellanea Mathematica*, Springer-Verlag, Berlin.

Reuleaux, Franz (1876), *The Kinematics of Machinery*, A. Kennedy, trans., MacMillan, New York.

Rosenbaum, Robert A. (1963), *Introduction to Projective Geometry and Modern Algebra*, Addison-Wesley, New York.

Rowe, David E. (2018), *A Richer Picture of Mathematics*, Springer, Cham.

Schopenhauer, Arthur (1909), *The World as Will and Idea*, 2 vols. R. B. Haldane and J. Kemp, trans., Kegan Paul, London.

Schwartz, Laurent (2000), *A Mathematician Grappling with His Century*, Leila Schenps, trans., Birkhauser Verlag, Basel.

Scott, Charlotte Angas (1900), "The Status of Imaginaries in Pure Geometry," *Bulletin of the American Mathematical Society* 6.

Shaw, James Byrnie (1918), *Lectures on the Philosophy of Mathematics*, Open Court, Chicago.

Smith, Stephen B. (1973), *The Great Mental Calculators*, Columbia, New York.

Steiner, Jakob (1832), *Systematische Entwicklung Der Abhängigkeit Geometrischer Gestalten Von Einander*, F. Fincke, Berlin, cited in (Moritz 1914).

Sylvester, J. J. (1870), "A Plea for the Mathematician," *Nature* 6(1).

Tao, Terence (2010), "A Type Diagram for Function Spaces," in *Compactness and Contradiction*, American Mathematical Society, Providence.

Thomson, William, and Tait, P. G. (1912), *Treatise on Natural Philosophy*, 2 vols., Cambridge University Press, Cambridge.

Veblen, O., and Young, J. D. (1910), *Projective Geometry*, 2 vols., Ginn and Company, Boston.

Venn, John (1894), *Symbolic Logic*, MacMillan and Co., London.

Weyl, Hermann (1939), "Invariants," *Duke Mathematical Journal*, 5(3).

(1955), *The Concept of a Riemann Surface*, 2nd ed., Gerald R. MacLane, trans., Addison-Wesley, Reading.

(2002), "Topology and Abstract Algebra as Two Roads of Mathematical Comprehension," in A. Shenitzer and J. Stillwell, eds., *Mathematical Evolutions*, Mathematical Association of America, Providence.

Wilson, Mark (2006), *Wandering Significance*, Oxford University Press, Oxford.

(2010), "Frege's Mathematical Setting," in M. Potter and T. Ricketts, ed., *The Cambridge Companion to Frege*, Cambridge University Press, Cambridge.

(2017), "A Second Pilgrim's Progress," in *Physics Avoidance*, Oxford University Press, Oxford.

(2020), "Review of Otávio Bueno and Steven French, Applying Mathematics," *Philosophical Review*, forthcoming.

Wittgenstein, Ludwig (1964), *Remarks on the Foundations of Mathematics*, E. Anscombe, trans., Blackwell, Oxford.

(1969), *On Certainty*, Denis Paul and G. E. M. Anscombe, trans., Blackwell, Oxford.

Wolf, Emil (2007), *Introduction to the Theory of Coherence and Polarization of Light*, Cambridge University Press, Cambridge.

Yang, A. T. (1974), "Calculus of Screws," in W. R. Spillers, ed., *Basic Questions of Design Theory*, North-Holland Publishing, Amsterdam.

Cambridge Elements $\equiv$

The Philosophy of Mathematics

Penelope Rush
University of Tasmania
From the time Penny Rush completed her thesis in the philosophy of mathematics (2005), she has worked continuously on themes around the realism/anti-realism divide and the nature of mathematics. Her edited collection *The Metaphysics of Logic* (Cambridge University Press, 2014), and forthcoming essay "Metaphysical Optimism" (*Philosophy Supplement*), highlight a particular interest in the idea of reality itself and curiosity and respect as important philosophical methodologies.

Stewart Shapiro
The Ohio State University
Stewart Shapiro is the O'Donnell Professor of Philosophy at The Ohio State University, a Distinguished Visiting Professor at the University of Connecticut, and a Professorial Fellow at the University of Oslo. His major works include *Foundations without Foundationalism* (1991), *Philosophy of Mathematics: Structure and Ontology* (1997), *Vagueness in Context* (2006), and *Varieties of Logic* (2014). He has taught courses in logic, philosophy of mathematics, metaphysics, epistemology, philosophy of religion, Jewish philosophy, social and political philosophy, and medical ethics.

About the Series
This Cambridge Elements series provides an extensive overview of the philosophy of mathematics in its many and varied forms. Distinguished authors will provide an up-to-date summary of the results of current research in their fields and give their own take on what they believe are the most significant debates influencing research, drawing original conclusions.

Cambridge Elements ≡

The Philosophy of Mathematics

Elements in the Series

Printed in the United States
By Bookmasters